Freedom Foods

Superior new Foods from low on the Food Chain for People, Producers and Our Planet

Mark Edwards

Eat Healthy. Eat Hearty.

Leave the e-footprint of a Butterfly.

The Green Algae Strategy Series

www.AlgaeAliance.com

www.AlgaeCompetition.com

www.FreedomFoodsNow.com

Key words.

Food	Microfarms	Sustainability	Fertilizer
Water	Nutrient recovery	Ecosystems	Hunger
Regenerative	Organic farming	Smartcultures	Poverty
Agriculture	Climate change	Energy	Drought
Aquifers	Micronutrients	Environment	Soil
Soil crust	Global awareness	Algaculture	Malnutrition
Genetics	Renewable energy	Biotechnology	Pollution
Microalgae	Industrial farming	Ecology	Algae

ISBN-13: 978-1461158585
ISBN-10: 1461158583
BISAC: Science / Biotechnology

Contents

Dedication

To the late Stuart Powrie, my extraordinary U.S. Naval Academy roommate and friend, who dedicated his life to preserving our freedom. While serving as a Navy Blue Angel, Lt Cmdr. Stuart Powrie made the ultimate sacrifice.

And to Stu's and all of our children, that they may avoid the horrors of war over food and the fossil inputs for food.

Forward

Freedom Foods offer a path forward with practical solutions for the critical food and energy challenges we face. This new food production system uses the innovative methods of abundance agriculture to move beyond the fossil resource dependent industrial agriculture. Freedom foods engage local growers in sustainable and affordable food and energy production.

Cultivating microorganisms for food, such as microscopic algae, offers significant productivity advantages over land-based crops. As these foods grow, they more efficiently use inputs such as water, land, energy, and nutrients. With some development, this method will enable growers to produce food locally, affordably and sustainably.

Freedom Foods offers new food supply with abundant agriculture that leverages algae production systems. We are at the beginning of an industry. We are establishing the principles of Freedom Foods. Some criteria established for Freedom Foods have already been achieved by algae producers or will be realized in the near future.

Algae cultivation is an evolutionary advance in agriculture

Over thousands of years, humans have increased food productivity with new inventions and growing systems. Each leap forward, up to now, has been accompanied by greater environmental costs.

Domesticating plants and animals encouraged the first permanent human settlements. Then, about 7000 years ago, irrigation brought water to the land and food surpluses supported the first great river valley civilizations. Thousands of years later, when the land salted up from over-irrigation, these civilizations vanished.

About 1000 years ago, the invention of an efficient plough in Europe allowed easier tilling of the soil. Europeans brought new areas under cultivation and new prosperity by cutting down Europe's vast forests.

About 100 years ago, the 19th century industrial revolution introduced mechanized agriculture, climaxing 50 years ago with the so-called 'Green Revolution' exported from the United States in the 1960s and 1970s.

Modern agriculture raised short-term productivity by using seed hybrids and massive fertilizer, pesticide, water and energy inputs. Higher productivity and cheaper foods have ignored hidden costs, such as consumption of non-renewable fossil fuels for fertilizers and machinery, pollution of soil and water through excessive use of chemical fertilizers, and depletion of soil and water, an ecological cost that will be borne by future generations.

Today, food and grain prices are rising dramatically around the world. The recent price spikes have not primarily been caused by disruptive weather events but are driven by long-term trends like growing population, climatic temperature increases, water shortages and competing land use for biofuel production. These trends make it more difficult to increase conventional food production, and food prices are going up, leading to riots and political instability as witnessed recently in the Middle East.

This is an inflection point in history when algae production, with its tremendous productivity and ecological advantages, moves from a niche activity to the main stage. Also about 50 years ago, algae were rediscovered as a possible new food. Just 30 years ago, algae burst into public awareness as a powerful new food with a promise as a food source to help feed the world's people. The algae industry is still in its infancy as a source of food, feed, medicine, biochemical and energy.

Algae food production is a productivity and ecological breakthrough

An algae production system can be an environmentally sound green food machine. Biomass can double every 2 to 5 days. On a given piece of land, growers can produce 40 times protein than corn and 400 times more protein than beef. Algae can flourish in ponds of brackish or alkaline water on unfertile land. In this way, it can augment the food supply not by clearing the disappearing rainforests, but by cultivating the expanding deserts.

Compared to the thousands of years of human food development, 30 years is only an instant in time. Cultivation of grains and development of irrigation took thousands of years. Soybeans, a newcomer, took 50 years to emerge from obscurity. The last 30 years progress in algae technology is remarkable.

Successful algae cultivation requires a more ecological approach than industrial agriculture. An algae pond is a living culture and the whole system must be considered. If one factor changes, the entire pond environment changes – quickly. Because algae grows so fast, the result can be seen in hours or days, not seasons or years like conventional agriculture. Instead of using conventional agriculture poisons, algae scientists balance pond ecology to keep out weed algae and zooplankton algae eaters without using pesticides or herbicides. Algae cultivation is a new addition to ecological food production.

Algae, the original life form on this planet, combined with new methods of cultivation, offers humanity the next productive food solution at this moment in history. Some envision huge centralized algae farms that will produce food and energy on a vast scale. With emerging technology, others envision a distributed model of smaller networked family and community algae microfarms, growing algae for valuable food and biomedicines for the nearby region.

Local food production avoids the increasing cost of transportation fuels. Local production redistributes the profits currently extracted by remote corporations all along the value chain in the multi-level food distribution system. A higher portion of the value of food sold to the consumer is returned to food producers, rewarding them and creating greater income equality and local self-sufficiency, for a more just and stable social fabric.

Algae microfarms for families and communities are coming soon

Over the past 30 years, many people have asked how they can grow algae themselves, in their own back yard. In fact, small-scale algae farming has been tested for 30 years all over the world, and models will soon be introduced as effective algae growing systems. One example of success is in France where a cooperative of 50 spirulina microfarmers support each other's success. They have created a school curriculum for growing algae. Growers are producing their own products and selling directly in their local region.

Soon, remote sensing devices linked with cell phone apps will assist the basic functions of algae culture health monitoring and diagnosis. This will allow local algae growers to consult with remotely located algae experts on how to maintain a healthy algae culture in their small production systems.

Growing food in cities and urban areas may become critical as fuel costs rise, making transported food increasingly expensive. On a small land area, a community could meet a portion of its food requirements from microalgae, freeing cropland for community recreation or reforestation.

Unlike plans for algae biofuels that require mega farms, algae foods can be produced on a small scale. As new technologies and systems design arrive, algae microfarming will be less costly. Progress will make growing easier and more accessible for more people around the world. Ecological communities can combine algae production with aquaculture and organic community gardens.

Microscopic algae are essential for individual and planetary health and restoration. The oldest photosynthetic life form is back. It represents a return to the origins of life. Algae and other microorganisms are the basis of Freedom Foods, offering us a new paradigm for abundance and security.

Please contribute your ideas, critiques, and extensions by engaging in the Freedom Foods Now open source collaboratory.

www.AgaeCompetition.com   www.AlgaeAlliance.com,
www. FreedomFoodsNow.com.

Robert Henrikson

Chapter 1. What are Freedom Foods?

Imagine superior fossil-free foods that clean rather than pollute our ecosystems.

Freedom foods reinvent our food supply from the foundation of the food chain. These foods give consumers the freedom to make smart choices to enhance health and vitality for people, producers and our planet. Foods grown low on the food web, microorganisms, require a fraction of the energy consumed by modern food. They are free of the resource consumption and waste caused by industrial crops because they are grown organically, with no or minimal fossil resources.

This alternative food supply provides splendid natural produce and does not compete with industrial agriculture because growers use abundance methods.[1] Growers recycle waste stream nutrients that are plentiful, affordable and will not run out. Scientific evidence and common sense show that food low on the food chain consumes fewer resources, while providing healthier nourishment for consumers, farmers, animals, and our ecosystems. These foods offer superior nutrition and taste with 100 times less pollution and waste.

Freedom foods are clean, healthy, and low in fat and cholesterol. They offer nutralence, high nutrient density that modern industrial foods are missing. These foods offer twice the protein of food grains, as well as 50 to 300% more nutralence, higher levels of essential vitamins, minerals, and antioxidants. Instead of polluting ecosystems, freedom food production repairs ecosystems, making them more resilient. They enable growers and consumers to leave the ecological footprint of a butterfly.

Freedom foods come from the most prolific plant on Earth, algae. Algae, and the other microorganisms algae attract, grow protein and other nutrients 70 times more productively than land plants, such as corn. Algae flour can substitute for corn, wheat, rice, soy or other food grain.[2] Algae offer a wide spectrum advantages because they are tiny and provide a dense nutrient package that is easily absorbed by the body. Animal feed produces superior meat, poultry and dairy products. These foods provide twice the nutritional density and 10 – 50 times more natural biodiversity than modern fossil-based foods.

Freedom to choose healthy food

The Centers for Disease Control published the Modified Retail Food Environment Index in April 2011 that shows that 9 out of 10 families lack access to retailers that sell healthy foods.[3] The Index reflects consumer access to retailers with fresh fruits and vegetables. Based on a range from zero (no food retailers that typically sell healthy food) to 100 (only food retailers that sell healthy food), the national average score was 10. Freedom foods can change the access problem by providing local foods that are fresh and healthy because these foods can be produced almost anywhere.

Modern consumers cannot currently make healthier food choices with freedom foods because these foods are not on the market yet. The USDA spends billions of dollars on fossil food research and subsidies, but largely ignores natural foods. Less than 1% of the USDA R&D budget supports organic production. The U.S. government has modest investments in algae research recently, but the focus has been biofuels, not sustainable and affordable food.

The USDA, FDA, and EPA have failed to require labels on genetically engineered, (GE) U.S. foods. Lack of food labels combined with weak enforcement has enabled "GE creep." In slightly over a decade, GE foods have gone from zero to where today they make up a major portion of packaged foods. Over 90% of U.S. food grains such as corn, wheat, and soy grow in GE monocultures that are refined into products loaded with fat, cholesterol, and calories devoid of essential nutrients, (empty calories). The USDA has approved 81 GE crops, while failing to deny a single proposal.[4] Applications pending propose to use transgenics to alter up to 30 genes simultaneously for a single crop.

Modern processed foods are high in fat and cholesterol. They cause obesity, diabetes, and a litany of Western diseases, including heart disease and cancers. Childhood obesity has increased 30% of the last 30 years and leads to diabetes. Diabetes is the leading cause of kidney failure, lower-limb amputations, and new cases of blindness in the U.S. Diabetes is a major cause of heart disease and stroke and the seventh leading cause of death in the U.S.[5]

Freedom foods are healthy and can treat and in some cases prevent obesity, diabetes, and other diseases. These foods are naturally biodiverse, which eliminates the need for GE monocultures. Unlike fossil foods, freedom foods are clean, free of chemical fertilizer and pesticide, herbicide, and fungicide residue.

Natural resource efficiency

Our modern food supply consumes massive amounts of resources and then pollutes our air, water and soil. Research suggests that around 50% of air pollution and over 60% of water pollution occurs from agricultural production.[6] The situation for soil loss is even worse because farmers over the last 40 years were forced to abandon 33% of the cropland globally due to industrial agricultural practices.[7]

The fossil resources required to produce industrial foods will be gone in a few generations. Long before the fossil resources become extinct, they will become unavailable or unaffordable. Fossil resources are already unaffordable to many farmers in India, China and Africa.

Organic production is healthier for people and producers than industrial, but represents less than 3% U.S. of cropland.[8] Organic farmers avoid, to the degree possible, GE seeds, chemical fertilizers and agricultural chemicals and poisons. Unfortunately, organic farming methods cannot meet global food needs because organic production uses more net fossil resources than industrial agriculture – fertile soil, freshwater, and fossil fuels. Organic production offers some ecological footprint advantages but nets about the same ecological footprint as industrial foods. Organic production has no more weather tolerance than other forms of agriculture.

A typical acre of corn raised with industrial methods in Michigan produces about 150 bushels of corn. Since corn contains about 23% protein, the acre produces about 200 pounds of protein. Input requirements include approximately 5 gallons of diesel fuel, 28,000 GE seeds; 150 pounds of nitrogen, 55 pounds of phosphorus, and 85 pounds of potassium fertilizers; a gallon of herbicide, and a gallon of pesticide/fungicide.[9] Irrigated cropland consumes three acre-feet of freshwater, about a million gallons.

Each of the 93 million acres of corn planted in 2011 in the U.S. adds 2.25 tons of CO_2 to the atmosphere and erodes 6 tons of soil loaded with waste fertilizer and poisons. The 2011 rains and floods multiply soil erosion over average years. The dead zone caused by agricultural run-off at the mouth of the Mississippi River in 2011 will be larger than the state of New Jersey.[10]

Each pound of algae, the foundation plant in freedom foods, contains about 50% protein; twice the level found in corn. An acre of algae production can produce 40 times more protein per acre than corn. Freedom foods grow without the fossil resources or GE seeds. Growers can use non-potable water, which saves freshwater for other purposes. Production does not use cropland and causes no erosion. Rather than polluting the air with carbon, each ton of algae sequesters or recycles 1.8 tons of CO_2. The only thing released to the atmosphere is a little water vapor from evaporation and lots of pure oxygen.

Climate chaos degrades or destroys fossil food yields and puts our entire food supply in jeopardy. Freedom food growers can produce consistently high yields, local to consumers, independent of weather, climate, altitude, latitude, geography or politics.

Mimic nature

Freedom foods mimic nature, exploiting the oldest, most efficient, and healthiest food production system on Earth. Algae and the diverse microorganism communities algae attract provide superior nutrition for people, animals, and plants.

Growers leverage the natural process that produces 70% of our oxygen and 40% our global biomass daily. Growers benefit from 3.7 billion years of nature's genius in sustained production by recycling energy and nutrients to provide efficient foods with practically no waste. Most the algae biomass offers value with less than 10% waste – which is typically recycled. Rather than following the fossil foods model that systemically degrades and pollutes ecosystems, growers clean and regenerate ecosystems.

Growers use abundant agricultural methods where they use plentiful resources that will not run out – sunshine, CO_2 and non-potable water.[11] These resources are typically free, surplus and affordable, which enables just about anyone to produce food. Growers recycle and reuse waste stream energy and nutrients. Of course, safeguards are necessary to assure the demise of waste stream parasites and pathogens. Growers use simple solar heaters or other technology that has been effective for over 40 years old.

A freedom food tortilla offers several times more nutrients, vitamins and antioxidants than a wheat, rice or corn tortilla. A hector of freedom food grows more protein and other nutrients in one year than a hector of corn produces in 40 years.

Corn produces its first gram of protein in about 120 days, a full growing season. Consumers and growers must wait another 365 days for the next harvest. Freedom foods growers produce the first gram of protein in about two weeks. Growers then harvest additional protein every few days, all year round. Freedom foods grow so fast, growers often harvest half the biomass daily.

Microflora communities are similar to the organisms in our gut that break down food and aid digestion. Growers cultivate microflora communities in microfarms and train them to produce food, feed, fertilizer, and many other coproducts. These tiny biofactories run efficiently as they recycle and reuse the residual energy and nutrients in the farm waste stream. Microfarmers using abundance methods produce good food while cleaning air and water and regenerating degraded soil.[12]

Higher nutralence

Algae provide a low fat, low calorie, nearly cholesterol-free source of protein. Some algae, such as spirulina, contain up to 70% protein by dry weight – twice the protein of meat. Unlike meat, most algae varieties provide the full complement of nine essential amino acids. The low fat content, only 5-10%, is a fraction of other protein sources. A chicken egg contains about 300 mg of cholesterol and 80 calories while providing the same protein as a tablespoon of the algae spirulina, which carries 1.3 mg of cholesterol and 36 calories. Algae are also an excellent plant source of glutamic acid, an amino acid that promotes intestinal health and immune function.

Each kilogram of algae biomass has roughly double the protein available from a kilogram of food grain. Algae concentrate many other nutrients at a multiple of the nutrients found in grains.

Growers have access to splendid natural diversity that enables them to grow biomass of 30 to 70% protein, depending on their target food, feed, fertilizer or coproduct. Growers that want to maximize lipids (oils) may select a species that contains 40% lipids. Other growers may want to maximize production of carbohydrates, pigments, vitamins, minerals, antioxidants, cosmetics, medicines, vaccines or many other valuable coproducts.

Algae absorb a wealth of mineral elements that concentrate as about one third of its dry biomass. The macronutrients include sodium, calcium, magnesium, potassium, chlorine, sulfur and phosphorus while the micronutrients include iodine, iron, zinc, copper, selenium, molybdenum, fluoride, manganese, boron, nickel and cobalt.

Note: Freedom foods include the spectrum of microorganisms such as algae, yeast, fungi, bacteria, archaea, protists, plankton, and others. The focus here is on algae, but a diversity of microflora may yield similar good foods with attributes superior to fossil foods.

Succulence is the natural ability of succulent plants to absorb and hold water.[13] Algae demonstrate "nutralence," as the biomass concentrates nutrients at substantially higher levels than land plants.[14] The reach of their roots and the nutrients available in the soil limits the nutrient density of land plants. Algae avoid that problem by living without roots. One tablespoon, 10 grams of algae delivers the same amount of:

- Calcium as 8 tbs milk, 32 tbs soybeans, 8 carrots, or 22 tomatoes.
- Magnesium as 40 tbs milk, 8 tbs soybeans, 9 carrots, or 6 tomatoes.
- Iron as 512 tbs milk, 8 tbs soybeans, 11 carrots, or 5 tomatoes.

Field studies show that algae nutralence, other vitamins and minerals are similarly 100 to 300% denser than field crop produce.[15]

Although very low in fat, algae offer an excellent source of the essential polyunsaturated fatty acids. The omega-6 and omega-3 fatty acids (ARA and EPA/DHA respectively) are necessary for normal metabolism, as they are the precursors to critical hormone-like, signaling molecules known as the eicosanoids. These short-lived messengers direct life-supporting functions such as blood clotting, inflammation, vasodilation, blood pressure and immune function. Only small amounts of ARA and EPA/DHA are needed daily (<1 g), and one tablespoon of algae can supply about half this amount.[16]

The fibrous components of algae add bulk to the digestive tract reducing hunger pangs, transit time, and intestinal pathologies.[17] The total fiber content of algae (~6 g/100g) is greater than that of fruits and vegetables promoted today for fiber content: prunes (2.4 g), cabbage (2.9 g), apples (2.0 g), and brown rice (3.8 g).[18] Algae's tiny cell size enhances bioavailability – absorption.

Health benefits

Phytic acid compromises the mineral availability from land plants, particularly legumes and grains, because the acid binds the minerals, rendering them unavailable for absorption into the blood stream. Phytic acid is typically absent in many algae species. Studies show that iron absorption is 3.5 fold greater for algae compared to rice.[19] Algae iron is easily absorbed by the human body because its blue pigment, phycocyanin, forms soluble complexes with iron and other minerals during digestion making iron more bioavailable. Hence, unlike iron derived from land plants, the bioavailability of algae iron is comparable to that of heme iron in meats.[20]

Algae contain a wide spectrum of prophylactic and therapeutic factors that include vitamins, minerals, amino acids and essential fatty acids. Algae provide the super anti-oxidants such as β-carotene, vitamins A, B, B-complex, C, D, E, and K, and a number of unexplored bioactive compounds.[21] These constituents stimulate numerous metabolic pathways and promote antioxidant, anti-bacterial, antiviral, anticancer, anti-inflammatory, anti-allergic, and anti-diabetic actions. Extensive medical research shows algae constituents, promote vascular, mental, and intestinal health.[22]

Algae nutrients, vitamins and minerals enhance physiological systems including the cardiovascular, respiratory and the nervous systems.[23] Algal components also activate the cellular immune system including T-cells, macrophages, B-cells and anti-cancer natural killer cells.[24] Algal polysaccharides inhibit replication of several enveloped deadly viruses including herpes simplex, influenza, measles, mumps, human cytomegalovirus, SARS, and HIV-1.[25]

Algae's nutralence, antioxidants, enzymes and extracts, boost the immune system and enhance the body's ability to grow new blood cells. Algae are rich in phytonutrients and functional nutrients that activate digestive and immune systems. Algae compounds accelerate production of the humoral system (antibodies and cytokines), enabling the body to protect against invading germs.[26] Specific algae polysaccharides have demonstrated anti-atherosclerotic functions, reducing blood LDL cholesterol concentrations, and cardiovascular disease risk.

Research on humans and animals shows algae components have utility in the prevention and control of diabetes.[27] Other studies have demonstrated algae's therapeutic value for cholesterol management, blood pressure, heart disease and cancers.[28] Algae moderate chronic inflammation, which often precedes degenerative diseases, diabetes and difficulties with fat metabolism.[29]

Research on mice shows algae delay the onset of motor symptoms and disease progression in ALS (Lou Gehrig's disease), reducing inflammatory markers and motor neuron death.[30] Algae are calcium rich and may protect against osteoporosis.[31] Recent research suggests algae activate human stem cells, which provide a spectrum of health benefits, including moderation of brain degeneration.[32]

Algae in human food history

Algae played pivotal roles in human evolution and survival. Early human societies evolved along coastlines, rivers and lakes and depended on algae for food and medicines.[33] The nutrient rich biomass was plentiful year-round and easy to harvest. Many groups ate algae directly and probably ingested algae in their drinking water.[34] Algae give water a sweet taste that would have been very attractive when the early *Homo* diet contained predominately dry, hard, bitter and sour tastes.

Algae provided a rich and nearly complete source of nutrition – a complex blend of nutrients that no other food source, plant or animal, could offer.[35] Algae were analogous to a modern-day vitamin supplement – but actually, algae are a more robust, natural, and inclusive blend of healthful nutrition.[36] Algae are a superior protein source, particularly the red, green and blue-green algae, which are as high as 70% protein (dry weight) which is higher than soybean (36%) and corn (23%).[37] Algae protein content is highest in the late winter/early spring, which is advantageous when terrestrial plant food sources are scarce. Algae nutralence benefited our ancestors year round.

Algae's rich savory taste

Sweetness dominates the human palate today, as illustrated by modern convenience foods. The human tongue has a fifth taste receptor, umami (savory or hearty), which would have been available primarily from algae in early hominid diets. The unique taste comes from three proteinogenic amino acids: glutamic, inosinic and guanylic.

Algae synthesize the hearty umami taste. Algae feeders such as fin and shellfish concentrate the savory taste that would have made these foods favored by early hominids. Today milk, aged cheese, and some meat products offer the umami taste.

The attractive savory taste of algae may have sparked brain enlargement in early *Homo* because algae provide the critical long chained fatty acids needed for brain growth and development.[38] Larger brains differentiated our ancestors from their cousins and enabled higher cognitive skills that aided survival.

Algae produce the rich umami taste with glutamate, which plays a key role in human cellular metabolism and digestion.[39] Digestion breaks down proteins into amino acids, which serve as metabolic fuel for other functional roles in the body. Glutamate is the most abundant excitatory neurotransmitter in the vertebrate nervous system and regulates several brain functions. Glutamate's role in body and brain functions is so critical that the logical explanation for the umami taste bud, called the mGluR4 receptor, was to attract our ancestors to glutamate. Algae are also an excellent plant source of glutamic acid, an amino acid that promotes intestinal health and immune function.

Many people assume algae lives up to its role expressed by uninformed journalists who repeatedly use terms such as slick, slime, scum, and other terms with negative attributions. Ironically, just the opposite is true. Algae's attractive taste may have helped us become human by attracting our ancestors to algae and the Omega-3s that sparked brain enlargement.[40] When some edible algae species are heated, the aroma is similar to chocolate chip cookies.

Convenience food?

Algae probably provided our ancestors with the original convenience food. Terrestrial foods were dry, hard, bitter and starchy. Land plants were difficult and risky to gather due to stealth predators. Algae offered a fresh, soft, delicious taste and were easily accessible. In many locations, algae were harvestable year round, which would have been a tremendous advantage when terrestrial crops were dormant or not producing.

What are Freedom Foods?

Early humans probably rubbed algae oil on their skin for sun protection. Algae add moisture and speeds the recovery from wounds, burns and bruises. Algae's high antioxidant activity protects skin from inflammatory reactions and sun damage.[41] Pacific Rim societies have been using algae for natural foods and remedies for centuries because they are effective. Organic chemists, medicinal chemists, biologists, and pharmacists are currently developing new anti-inflammatory and anti-cancer medicines from algae.[42]

As early humans migrated out of Africa, they followed coastlines where macroalgae – seaweeds and sea vegetables – were plentiful. At low tide, hominids could harvest algae easily and dry it quickly in the sun. The light produce probably served as the first wampum in trade because it was easily transportable. Algae wampum offered a side benefit; a hungry trader could eat the product.

Members of the Chinese Court, around 1,100, harvested and reserved a specific algae variety for the Chinese Emperor. The Japanese reserved another algae variety for the Samurai, the Japanese nation's fiercest warriors. Today, Chinese Olympic athletes consume algae daily because, like the Samurai, algae nutrients enable them to train harder and longer. The therapeutic elements provided by algae allow the athletes to recover from injuries faster.

The Aztecs used algae for food, medicine, trade and religious ceremonies. Indigenous people along coastlines or lakes have harvested natural stands of algae for millennia for use as food, feed, medicines and trade. Algae probably protected our ancestors against many diseases including scurvy, xerophthalmia (blindness from vitamin A deficiency), goiter, arthritis, diabetes, mental retardation and others. The Chinese have used algae for medical purposes for centuries because these natural remedies are safe and effective.

Grow low on the food chain

Consumers have two options when choosing to eat low on the food chain; algae or the animal and plants fed algae. Human consumers receive the most benefit from eating algae directly or algae-based foods. We also benefit from eating fish, fowl, dairy and meat animals that received some of their nutrition from feed low on the food web. We also benefit from field crops grown with organic algae biofertilizer because the produce offers significantly more nutrients than crops grown with chemical fertilizers.[43]

People benefit from eating low on the food chain when they eat:

- Algae directly, such as sea vegetables or other algae foods.
- Foods made from algae flour such as crackers, chips, dips, pasta, cakes, or cookies.
- Foods made from algae texturized vegetable protein such as alfu, (algae tofu) or vegetable meat products.
- Foods made with algae nutrients and other ingredients.
- Produce grown with algae biofertilizer in lieu of chemical fertilizers.
- Fish or fowl grown with algae feed or supplements.
- Meat or dairy animals raised with algae feed or supplements.

Some non-Asians shy from eating algae directly, except in soups and sushi. Most consumers are surprised that food processors already integrate algae components in our food supply. A market basket test at Arizona State University examined the non-fresh produce items in 10 typical shopping carts for foods containing algae components. About 72% of the foods and 88% of the cosmetics contained algae constituents.

Algae Flour Makes Delicious Freedom Foods

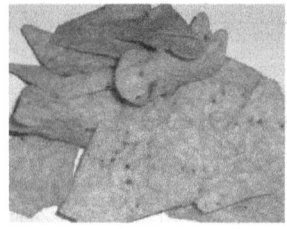  

Algae synthesize the omega-3s that fish accumulate in their oil and that support human brain, eye and heart functions. Algae-based foods provide omega-3s along with other essential nutrients. Per kilogram, these foods provide 10 times the beta-carotene as carrots and 10 times more vitamin B-12 and iron than beef liver.

Beer and soft drinks use algae as a clarifier. Algae cell walls contain carrageenan, used as a stabilizer or emulsifier commonly found in dairy, confectionary and bakery products. Alginates provide alginic acid from brown algae, which thicken liquids and make them creamier and more stable over wide variations in temperature, acidity, and time. About half of the alginate harvested today goes into ice cream and other dairy products to make them smoother and prevent ice crystals. Alginates also keep toothpaste and lipstick from going dry.

Algae contains considerable agar, a polysaccharide that solidifies almost any liquid. Agar acts as a colloidal agent used for thickening, suspending, and stabilizing soups, stews and canned goods. Agar is used as a substitute for gelatin, as an anti-drying agent in breads and pastries and also for gelling and thickening. Agar enhances processed cheese, mayonnaise, puddings, creams, jellies, and frozen dairy products.

Algae naturally form flours that can substitute for food grains such as rice, corn, wheat or barley. Algae flour makes chips, dips, breads, tortillas, crepes, cakes and pretzels that have superior nutrient profiles to their cousin foods with substantially lower fat and cholesterol. Of course, people will not choose the algae-based food models unless they taste similar or better and are affordable.

Algae Makes Delicious Freedom Foods

Algae milk can substitute for dairy, soy, or almond milk while providing higher protein. Many people are lactose intolerant and others are bothered by soy and nut allergies. Algae milk induces neither intolerance nor food allergies. Algae sugars can substitute for cane, beet or corn sugars. Solazyme announced new algae chocolate with 85% fewer calories.

Texturized algae made into vegie burgers pack the savory umami taste. Currently, the extremely popular Umami hamburger chain in Los Angeles must add MSG to their beef to gain the attractive hearty flavor. When the Umami firm uses algae, they will be able to provide great flavor with natural savory taste from algae.

Availability

Sea vegetables and other algae products are popular for their nutrition, color, taste and texture in most Asian Pacific Rim countries. Many high-end European restaurants offer an algae appetizer and entre. In the U.S., freedom foods are available only in Asian markets and health food stores, where they sell as food supplements. Whole Foods and Trader Vic's recently began selling algae snack foods. Sushi bars combine several forms of algae with other seafood and rice.

Food companies have investigated algae-based foods for decades but they faced a pair of showstoppers with freedom foods – supply and demand. Growers have no incentive to grow algae when terrestrial foods are so heavily subsidized and cheap. Consequently, food-processing companies have no suppliers.

General Foods, Nabisco, Frito-Lay, Borden, Dial or Quaker Oats cannot manufacture an algae product today because there is no market. Consumers do not select these foods because they know little about the health and environmental value of freedom foods.

The Freedom Foods Revolution plans to spark demand by educating consumers, which will motivate new suppliers. The new freedom foods industry will offer thousands of engaging entrepreneurial opportunities for growers, suppliers, restaurateurs and chefs.

Chapter 2. Food Freedom of Choice

We need a revolution to take back our liberty to choose healthy foods.

Freedom foods free people, producers, and our planet from the tyranny of big agribusiness and the insults their fossil foods and genetically identical monocultures exact on us. If a foreign country inflicted the damage on us levied by industrial agriculture, we would declare war on that country. Those who monopolize our farm policy and dictate what we have to eat – big agribusiness – have stolen our food liberty.

Industrial agriculture devours enormous levels of fossil inputs – genetically engineered, (GE) seeds, fertile soils, freshwater, fuels, fertilizers and agricultural chemicals and poisons. Industrial food inputs are non-renewable and current massive consumption is unsustainable. Dwindling supplies will drive up costs until they are unaffordable or unavailable; amplifying food insecurity.

When our children are hungry and need food, they will find the fossil resources to produce industrial foods are extinct and their ecosystems hopelessly polluted. What will our children do for food? They will need a fossil-free food supply.

Prince Charles has advocated sustainable agriculture with less dependence on fossil resources for over 30 years. He shared his insights in a keynote address at the Georgetown University Future of Food Conference in May 2011. He conveyed his motivation as:

> I have no intention of being confronted by my grandchildren, demanding to know why on earth we didn't do something about the many problems that existed when we knew what was going wrong. The threat of that question, the responsibility of it is precisely why I have gone on challenging the assumptions of our day. And I would urge you, if I may, to do the same because we need to face up to asking whether how we produce our food is actually fit for purpose in the very challenging circumstances of the twenty-first century.

> We must take care of the earth that sustains us because if we don't do that; if we do not work within nature's system, then nature will fail to be the durable continuously sustaining force she has always been. Only by safeguarding nature's resilience, can we hope to have a resilient form of food production and insure food security in the long term. This, then, is the challenge facing us.[44]

Prince Charles provided evidence of scarcity and unsustainability for each of the fossil resources required to produce industrial foods. Prince Charles advocates for genuinely sustainable agriculture for the long term that replenishes soil and water, is not dependent upon the use of chemical pesticides, fungicides and insecticides nor artificial fertilizers, growth promoters or GE monocultures. He noted that we must reduce the use of those substances that are dangerous and harmful, not only to human health but to the health of those natural systems such as the oceans, forests and wetlands.

Prince Charles recognizes that food production concentrated in large farms distant from consumers puts our food supply in jeopardy. A single weather or geological event could destroy large food supplies. In addition, with increasing fuel scarcity and cost, long distance food transportation will not be possible. He advocates a more resilient system were many smaller growers produce food close to consumers. He notes "...strengthening small farm production could be a major force in preserving the traditional knowledge and biodiversity that we lose at our peril."[45]

Food riots and revolutions

Over 40 nations endured food riots in 2008 due to food scarcity and affordability. Several of the 2011 revolutions in North Africa ignited because the dictators were unable or unwilling to supply their citizens with sufficient food. People in Tunisia and Egypt were tolerant of their tyrants until families could not find affordable food. Food riots exploded into revolution, which toppled governments and ousted the dictators.

Today, half the people on Earth are hungry and food insecure. As the fossil resources required to produce food become increasingly scarce, more people will suffer from malnutrition and hunger.[46] Unfortunately, more countries also will experience food riots or revolutions unless we change to sustainable and affordable food production methods.[47]

Fatal errors

A food supply based on fossil resources imposes five fatal errors on societies; constantly rising prices, freshwater scarcity, soil erosion and degradation, severe environmental insults and eventually crop failure.

Input prices. Food prices escalate in lockstep with the price of oil and other fossil resources because these resources are necessary for production. The price of a single fertilizer, phosphorous, rose 700% over a recent 14-month period. As natural resource scarcity increases, food costs will rise even faster because speculators will hoard resource assets. Speculators have already disrupted soil, oil, water and fertilizer markets.

After adopting industrial agricultural practices imported from the U.S., over eight million farmers in India quit farming during the 1990s due to rising crop input prices – seeds, water, fuels, fertilizers and chemicals. Rising input costs rapidly escalated farmer debt. In the decade ending in 2007, 183,000 farmers in India committed suicide because their farms could no longer provide for their families.[48]

In the U.S., about 30% of the working population farmed at the beginning of the green revolution. Today, the USDA reports that less than 1% of the population farm, largely because family farmers cannot afford the crop inputs. Family farms have been replaced with large farms and huge agribusiness corporate farms that hire farm managers that practice industrial agriculture.

Water scarcity. Agriculture consumes 70% of the freshwater globally and 80% in the western U.S.. Recent cropland expansion has converted deserts to fields with irrigation. The sunshine is great for crops but desert farming consumes tremendous amounts of freshwater.

Countries, states, cities and farmers have been fighting over water for decades. As global warming intensifies droughts and causes more water loss from evaporation, water conflicts will intensify. Vandana Shiva in *Water Wars: Privatization, Pollution and Profit* predicts the water wars of the 21st C may surpass the oil wars of the 20th C.

Soil degradation. Farm operators now lease over half the cropland in the U.S.[49] Non-owners are motivated to maximize short-term profitability and have less motivation than owners to use sustainable agricultural practices. Systemic extraction, waste, and pollution systemically degrade ecosystems until they become unproductive and must be abandoned.

Industrial agriculture promotes erosion by plowing fields with heavy equipment, intensive cultivation, and application of harsh fertilizers and agricultural chemicals. Growing crops constantly disrupts and compacts topsoil while extracting nutrients and humus, which amplifies erosion and wears out the soil. As fields deteriorate, farmers react by applying more water, fertilizers and pesticides. These actions not only increase costs but also make run-off more toxic.

Environmental insults. Agricultural pollution causes significant health problems that often lead to disability and premature death for farmers and people living in rural communities. People call the trains from the city of Chotia Khurd in northern India cancer trains because so many people in the farming villages must travel from their homes to the city for cancer treatments.[50]

China's farmers must abandon over a million acres of degraded cropland each year. The resulting dust storms are worse the U.S. Dust Bowl in the 1930s. China's air pollution caused over 20 million people to suffer respiratory illnesses in 2007. The country's health ministry demanded that the World Bank remove mortality calculations from a report on the country's air and water pollution because the numbers could trigger social unrest.[51]

Americans do not escape air pollution, largely due to agriculture. The American Lung Association's 2011 State of the Air reports that 155 million Americans, just over half the nation's population, live in areas where air pollution levels are often dangerous to breathe.[52]

Crop failure. When soil, oil, freshwater, phosphorus or any other nonrenewable input becomes unaffordable or extinct locally, the food supply crashes. Food supplies have crashed in many countries already. In the Mid-East, insufficient freshwater ended food production in many regions. In parts of Central America, Africa and China, erosion and expanding deserts have destroyed cropland. In India, the cost of fertilizer, water scarcity, and ecological pollution has forced farmers from their land. The serious threat hidden in the fossil foods model is that when just one of the fossil inputs becomes unavailable, the entire food supply fails.

New food supply

We could avoid the fatal errors with fossil foods, if we were able to design an alternative food supply based on abundant resources that do not rise with the oil, do not pollute and will not run out. We could transform our food production system so that consumers could make healthy choices for themselves, producers, our planet and our atmosphere, Figure 2.1.

This new food supply will enable us to leave a positive legacy for our children – healthy, affordable food, clean ecosystems, breathable air, and abundant natural resources. Freedom foods offer a clean, naturally biodiverse and healthy alternative to fossil foods. Growers produce foods and feeds low on the food chain with abundance methods that use minimal or no fossil resources to produce foods.

These foods can be grown locally or regionally and are superior in nutrition and taste yet create minimal pollution or waste. The use of plentiful resources that will not run out presents the opportunity for food democracy, where everyone has access to good food or the inputs to grow their own food.

Figure 2. 1 Freedom Food frees Consumers for Healthy Choices

People	Producers
• Healthy, affordable food choices	• Crops with affordable, non-fossil inputs
• High nutrient and vitamin density	• Grow fresh and local to consumers
• No empty calories or GE material	• Grow 70 times faster and fossil crops
• Organic and naturally biodiverse	• No fertilizer or pesticide residue
Planet	Atmosphere
• No waste or pollution	• Moderate climate chaos
• Regenerate fields and ecosystems	• End black soot pollution
• Stop natural resource extraction	• Capture greenhouse gasses
• Save biodiversity and stop extinctions	• Remove 50% of food trucks from roads

Freedom Foods

The Freedom Foods Revolution proposes to distribute the knowledge and capability for abundance methods globally to enable all people to grow good food and other coproducts for their family and community locally. Distributed, local production produces fresh foods that do not need preservatives and avoid the high dollar and energy cost of transportation. Abundance democratizes the access to food production and breaks the yoke of agribusiness tyranny.

People

Our food supply system should provide healthy, clean and nutritious food for people. Industrial farmers use predominately

GE seeds to produce monocultures that are refined to foods high in fat, sugar, salt and preservatives, but low in nutrients per calorie. Crops grown in nutrient deficient soil suffer from hidden hunger, and may have up to 75% less of the deficient nutrient in foods compared with fertile soil.[53]

Nutrient deficient plants may cannibalize their cell walls and storage nutrients to insure survival. Nutrient deficiencies are often visible in the field, but may not be apparent in the produce. Produce with nutrient deficiencies transfer hidden hunger to human consumers, which expresses in the many maladies and diseases caused by nutrient deficiencies.

Hidden hunger from nutrient deficiencies imposes a huge toll on society, according to the UN World Health Organization, (WHO).

- Vitamin and mineral deficiencies account for 10% of the global health burden – second only to clean water.
- Children and adults with micronutrient deficiencies suffer impaired development, disease and premature death.
- Over 2 million children die unnecessarily each year because they lack vitamin A, zinc or other nutrients.
- Over 18 million babies are born mentally impaired due to iodine deficiency each year.
- Iron deficiency undermines the health and energy of 40% of women in the developing world. Severe anemia kills more than 50,000 women a year during childbirth.[54]

Scientists use the term empty calories to describe how nutrient dilution from industrial farming diminishes the nutrients in each bite. Nutrient dilution occurs because farmers maximize yield weight, not food quality or nutritional density. Yield weight increases with GE foods, but most the extra weight comes from water. Nutrient dilution also depletes modern foods of color, aroma, taste, texture and nutrient density. GE monocultures jeopardize our entire food supply because a single pest vector can destroy the entire crop. GE monocultures also cheat consumers of nutralence, nutrient density, and diversity.

Our human bodies evolved over eons to eat nutritionally rich and diverse natural foods; not GE monocultures loaded with preservatives and chemical residues. GE crops require massive amounts agricultural chemicals and poisons because they are genetically engineered to optimize yield, not plant vitality. GE crops are extremely vulnerable to a broad spectrum of weeds and pests, so they require significantly more fertilizer and chemical poisons, some of which stays as residue in refined foods and fresh produce. Recent scientific evidence suggests that despite Monsanto's promise of GE food safety, genetic crop manipulation causes changes in human body chemistry.

Chemicals in fertilizers and pesticides have been linked to ADHD, autism, cancer, Lou Gehrig's disease, and other illnesses. The latest research, published in May 2011, as three independent studies in *Environmental Health Perspectives*, shows children exposed to pesticides in the womb are more likely to have measurable problems with intelligence, memory, and attention beginning at 12 months and continuing through early childhood.[55] These studies link prenatal pesticide exposure (measured in the urine of mothers-to-be) to significantly lower IQ in children by age 9. The research teams, from Mount Sinai School of Medicine, Columbia University's Mailman School of Public Health and the school of public health at the University of California, Berkeley, all conclude that pesticide exposure during pregnancy could negatively affect brain development.

Animal studies have previously demonstrated that organophosphates, (OP) scramble brain function and behavior in baby rats. Two studies in 2010 found that children exposed to higher levels of organophosphate pesticides than their peers were more likely to be diagnosed with attention deficit hyperactivity disorder, (ADHD).

One recent study followed hundreds of mostly Latino mothers and children in California's Salinas Valley, a center of commercial agriculture. Many of the women were farmworkers, or had family members who worked on farms. When the women were pregnant, the researchers tested their urine for several chemical by-products of organophosphates -- a standard means of gauging exposure. The mothers with the highest levels of by-products, known as metabolites, had children whose IQs at age 7 were seven points lower, on the 100-point scale), than the children whose mothers had the lowest levels of exposure.[56]

Leonardo Trasande and colleagues at the Mount Sinai School of Medicine estimated the annual cost of environmentally mediated diseases in U.S. children to be $76.6 billion in 2008.[57] Since roughly half of air pollution and a majority of water pollution come from industrial agriculture, the agriculture creates a cost to our children of at least $33 billion. Of course, medical costs ignore the loss of life quality and family disruptions caused by childhood cancer, asthma, intellectual disability, autism, and attention deficit disorder. No similar studies have examined the elderly but the costs would probably be higher since the elderly are a larger group and have substantially longer exposure to environmental pollution.

Producers

Fossil foods adversely affect growers too. Genetic engineers trick plants to transfer their limited energy from roots to fruit, which creates more food production by weight, but at a steep price for both consumers and growers. Much of the additional weight comes as water weight, which cheats consumers with additional nutrient dilution.

Farmers must pay monopolistic prices for their GE seeds and cannot save their seeds, as farmers have for eons. Farmers often have to pay more for their seeds each year. GE crops need protection from natural plants, which add to farmer cost and labor. Shorter roots require farmers to apply additional freshwater irrigation to retain soil moisture in the narrow root zone. Shorter roots limit the plant's reach for nutrients, so farmers must apply significantly more chemical fertilizers. GE plants cannot compete with natural weeds, so farmers have to apply additional agricultural poisons – pesticides, herbicides and fungicides. A recent study by the Union of Concerned Scientists revealed that in spite of Monsanto's claims that GE crops use fewer agricultural poisons, producers are actually using more pesticides and herbicides.[58]

Soil degradation and loss operate in a slow insidious manner that erodes farmers' ability to grow nutritious crops. Plant pathologist Don Huber, professor emeritus from Purdue notes that crops usually get enough phosphorus, potassium and other common minerals to grow, but often cannot draw sufficient micronutrients from the soil to fend off diseases.[59] Such nutrients include the metals manganese, copper, zinc, iron and boron. Crops deficient in micronutrients pass their hidden hunger on to animal or human consumers.

In order to maximize short-term production, industrial agriculture manipulates the natural ecosystem and disrupts, degrades and displaces precious soils and nutrients. Industrial agriculture accelerates erosion around 500 times faster than natural processes.[60] Soil carried away by erosion contains about three times more nutrients than are left on the remaining soil.[61] Bruce Wilkinson estimates that global erosion occurs at a rate of about 75 billion tons a year.[62] Moving these amounts of rock and soil would fill the Grand Canyon in Arizona in a generation.

Industrial agriculture erodes 1.8 billion tons of soil from U.S. cropland each year.[63] According to the EPA, agricultural pollution leaves over 52% of our waterways unfit for human recreation.[64] In some agricultural states, pollution makes over 90% of the waterways unfit for fishing.

Fossil foods place farmers and their neighbor's health in jeopardy from environmental pollution. Erosion not only removes topsoil, organics and nutrients, but also carries sediment to waterways along with agricultural poisons. Most crops absorb only about 5% of the pesticides applied and most of the residual enters wetlands, waterways and groundwater.

Global climate chaos

Industrial agriculture consumes nearly 20% of all fossil fuels and contributes 30% of the greenhouse gasses that cause global warming.[65] A single cow-calf combination produces more greenhouse gasses than a medium size car traveling 15,000 miles. Cultivation creates dust and promotes additional dust from wind, which adds tons of fine grain particulates to the atmosphere. Choking dust carrying agricultural chemicals occludes the sun and causes respiratory distress for millions of people.

Each hector of corn production creates an average of 18 tons of soil erosion, which carries enormous amounts of chemical fertilizers and agricultural chemicals.[66] Soil erosion first diminishes and degrades food production and then pollutes the ecosystems. The U.N. Food and Agricultural Organization, (FAO), reported that erosion and nutrient depletion forced farmers to abandon over 30% of the fertile soil globally over the last 40 years. In addition, our planet has over 400 dead zones, where agricultural pollution has killed all aquatic life. A recent study published in *Science* reported dead zones are growing globally at about 10% each decade.[67]

Bill Food Rights

The Freedom Foods Revolution proposes breaking the tyranny of big agribusiness and inventing a new food system that benefits people, producers and our planet. We propose a Food Bill of Rights and Protections that offers liberty to consumers and growers and benefits to our planet, Table 1.

Access to affordable good food represents a fundamental right and creates social equity. When people lack access to good food, they create a substantial drag on all of society. When all people have access to food, everyone can contribute to improving society, our foods, and our ecosystems.

Food security. People should have access to good food or the inputs for growing their own food. Over half our global population, 3.5 billion people are food insecure and often hungry.[68] Many of their children and elders are malnourished. A child in Africa dies of malnutrition every 5 seconds.[69]

Table 1. Food Bill of Food Rights
for People, Producers and our Planet

Food rights	Description
1. Food security – Food democracy	All people have access to sustainable, affordable, and delicious food.
2. Healthy choices	Natural and clean, nutrient dense, and low in fat and cholesterol. Cures rather than causes obesity, diabetes, and other diseases.
3. Spiritual and dietary choice	Consumers may choose foods aligned with their spirituality and diet values.
4. Excellent sensory appeal	Superior color, aroma, texture, and taste compared with fossil foods.
5. Creates local jobs	Offers good local jobs with a living wage.
6. Preserves natural resources	Fossil free and avoids extraction, which preserves natural resources for our children.
7. Fresh and local	Meets our 50/50 goal where 50% of food grows within 50 miles of consumers.
8. Naturally clean and biodiverse	Millions of natural species grow, which avoids the need for genetic monocultures.
9. Cleans and repairs ecosystems	Cleans and regenerates ecosystems so growers can leave every field better than they found it.
10. Climate independence	Reliably produces food independent of climate, weather, geography, or politics.

In the U.S., where we are blessed with ideal growing conditions, one out of five Americans – over 60 million citizens – receives food support because they are food insecure and hungry.[70] Food support, food stamps and school lunch programs, provides about $1 per meal. Dollar meals eliminate the opportunity for families and their children to participate in the American dream. Poor Americans lack food freedom. They must feed their children cheap, nutrient deficient foods that cause obesity, fatigue, dull brains and diabetes.

One out of every four American children struggles to get enough food for their bodies and minds to develop properly.[71] Food insecurity forces our children to eat cheap food, which typically delivers loads of fat, slat and cholesterol, but poor taste, color and calories devoid of nutrition. Malnourished children incur developmental impairments that limit their physical, intellectual and emotional development.[72]

The structure of our food supply should be a top priority for national defense, security, health and education. Politicians and policy leaders ignore the strategic value of our food supply – to our peril.

The heavy equipment and the physical labor requirements of industrial agriculture preclude large parts of the population from growing food. Women, the elderly and the disabled have near-zero opportunity to produce industrial foods, magnifying food injustice.

Healthy choices are not available to most consumers today because freedom foods are not yet widely available. Most of our industrial foods create rather than prevent health problems.

Modern foods lack informative labels. The European Union considers food labels the most important tool for ensuring the freedom of choice. The EU gives this freedom to know and freedom to choose by EU law. GE food products must clearly state the presence of GE food on the label. Clear labels give every consumer an opportunity to make an informed decision. Freedom to choose non-GE foods does not currently exist in the U.S. because the FDA does not yet require labels for GE crops. GE monocultures dominate the U.S. food supply, with 90% or more of our basic food ingredients coming from GE crops: soy, corn, canola, and sugar beets. Most modern packaged foods also contain chemical, pharmaceutical and pesticide residues.

Many consumers, especially children, must eat cheap foods high in fat, salt and calories but low on nutrition because they cannot afford healthy foods. Consumers deserve affordable healthy foods that promote strong minds and bodies. Foods should prevent rather than cause obesity, diabetes, heart disease, cancers, and the other Western diseases.

Our food supply should liberate our children from hidden hunger and the silent but costly nutrient deficiencies stemming from nutrient dilution. Fossil foods create nutrient dilution because the crops systemically extract soil micronutrients that farmers do not replace. (Farmers do replace the macronutrients in order to maximize yield.) Consumers deserve to know when nutrient dilution occurs in produce or processed foods. Current food labeling provides far too little information on nutrient availability, density, quality and diversity.

The Centers for Disease Control, (CDC) recently reported that one out of three American children born after the year 2000 will contract diabetes – predominantly due to a poor diet of nutrient-deficient calories.[73] Over 40% of women are likely to contract diabetes. The plague of obesity and diabetes creates havoc on our educational system and creates immense drag on our health system. The CDC estimate the total costs from diabetes at $174 billion annually.[74] Consumers should have access to healthy foods that prevent rather than cause diabetes.

Consumers expect clean, natural foods. Fossil food production requires heavy chemical fertilizer and pesticide applications to avoid natural competitors and to maximize yield weight. Most of the processed foods in the U.S. contain unnatural, GE plant material as well as synthetic preservatives and pigments.

The GE crops require farmers to apply huge amounts of agricultural chemicals and poisons, which imposes substantial health risk to farmers, their families and their farm animals. Aerial spraying drifts far from fields and flows into wetlands and waterways. Exposure to polluted ecosystems creates prolonged health risk for rural communities, diminishing their quality of life, and the value of their property.

Industrial farmers also experience substantial physical and health risk from working with heavy equipment, hard physical labor, and fatigue. Modern agriculture competes with fishing and mining for the industries with the highest rates of disabilities and death.

Spiritual and dietary choice. Our foods should enable people who choose a special diet or spiritual path such as Buddhism to enjoy good foods. Vegetarians and vegans have very limited good food choices with fossil food products.

Excellent sensory appeal. Nutrient dilution and hidden hunger in crops causes not only empty calories but also the degradation of color, taste, aroma and texture. A modern field tomato has practically no taste and poor color and texture due to modern food cultivation and harvesting practices. Consumers should have access to fresh whole foods with excellent sensory appeal.

Local jobs. The 322,000 principal U.S. farm operators, only 0.001% of the U.S. population, on 322,000 farms produce 90% of all foods and fibers consumed in the U.S., plus another 11% for export.[75] The USDA farm policies have practically decimated local family farms.

Government subsidy policies designed to support family farmers instead spawned a rich oligarchy at the top U.S. agricultural production. These few largest and wealthiest agribusiness scions orchestrate Congress to award continually more transfer payments in the form of subsidies, tax relief, trade barriers, foreign aid, and other incentives to enrich themselves. In contrast to political rhetoric, small farmers realize negligible benefits from government farm policies, and subsidies.

Concentration of power in big agribusinesses siphons off the wealth from family farms and concentrates money spent on food in the pockets of a few. Family farmers are unable to compete with huge producers who monopolize access for the fossil resources required to produce food. Large agribusinesses employ legions of poorly paid, unskilled labors that have sparse access to good education, housing, healthcare and healthy food.

Farm subsidies magnify inequity and waste billions. Over 70% of the $170 billion in farm subsidies over past 15 years supported the production of just five crops: corn, wheat, cotton, rice and soybean. Just four of those same favored five: corn, wheat, cotton, and soybean accounted for over 70% of the $25 billion in crop insurance over 15 years.[76]

From 1995-2009, the largest and wealthiest top 10% of farm program recipients received 74% of all farm subsidies with an average total payment over 15 years of $445,127 per recipient. The bottom 80% of farmers received an average total payment of just $8,682 per recipient.[77] Large agribusinesses and wealthy farmers game the system to gain the agricultural subsidies put in place to support small family farms. They use the wealth given by the government to buy land and other resources from smaller farmers, many of whom do not receive subsidies.

Subsidy policy should make agriculture cleaner, healthier, more sustainable, and more fair and equitable for the diverse interests in the U.S. food and fiber system. Current agricultural subsidies amplify social inequity, enrich those who are already wealthy, and promote only a few crops. The current subsidy policy rewards and encourages actions counter to the Food Bill of Rights.

Canada, Mexico and other close allies have outstanding lawsuits against U.S. subsidies with the World Trade Organization because these subsidies substantially depress the real price of food grains. British International Development Secretary Douglas Alexander said "It's unacceptable that rich countries subsidize farming at $1 billion a day, costing poor farmers in developing countries $100 billion a year in lost income."[78] Subsidies hide the true cost of food, create incentives for unsustainable farming practices and bloat the financial deficit. We need to transform subsidies so they motivate production of environmentally sustainable crops that repair ecosystems.[79]

The World Trade Organization ruled that U.S. cotton subsidies, contravene international rules. The U.S. government "settled" the suit in 2010 by agreeing to subsidize Brazil's cotton farmers over half-billion dollars over the next several years.[80] WTO subsidy lawsuits from Canada, Mexico and several African countries remain unresolved.

Growers and the U.S. economy will benefit from a food system that transforms crop subsidies and creates good local jobs that provide a living wage.

Preserves natural resources. Industrial food production consumes massive amounts of increasingly scarce fossil resources. Each year the cost of fossil resources increases, pushing up food prices. The price of GE seeds doubled last year in some markets. GE crops accelerate fossil resource depletion because they require significantly more fossil inputs than heritage crops.

Most good cropland has been farmed for generations. Additional land is less flat and fertile and requires more of every input, making food production continually more expensive.

Farmers face scarcity and higher costs for freshwater. Globally, the common metric for food grains is 1:1,000; one ton of grain consumes 1,000 tons of fresh water.[81] Water consumes energy because roughly 30% of the stationary energy used in the U.S. goes to move water for irrigation.[82] Few people, politicians or policy makers are aware of the tremendous energy costs in moving irrigation water because water receives high subsidies in the U.S. and the cost data is typically not publically available.[83] In some areas, farmers pay less than 2% of the true cost of irrigation water. When subsidies make a public resource nearly free, users have little motivation to conserve.

Practically every food producing country faces freshwater scarcity. Cities are claiming freshwater that had been allocated to farms. Scientists predict many of the fossil aquifers on which food supplies depend will crash within the current generation. Farmers will have to abandon the land when the aquifers go dry because global warming creates more heat and droughts than rain. Climate chaos brings more severe storms that cause floods and erosion. Floods do not recharge aquifers effectively because the water moves too fast to percolate into the soil.

Farmers consume billions of gallons of fossil fuels in cultivation and harvest. Food processing, packaging and transportation take additional billions of gallons. Fossil foods require enormous amounts of fuel-intensive fertilizers and agricultural chemicals that must be mined, processed, shipped and applied to fields. Farmers also apply millions of tons of poisons made with fossil fuels. A food supply that avoids fossil inputs will preserve resources that would otherwise be extinct when our children need them.

Modern food production forces consumers and growers to leave mammoth carbon and ecological footprints that accelerate global warming and climate chaos. The average American foodstuff travels an estimated 1,500 miles before being consumed.[84]

Fresh and local. Consumers deserve access to fresh foods grown locally, which avoid chemical preservatives and synthetic dyes. Local foods can reduce up to 80% of food costs by eliminating most the refining, preservatives, packaging, storage, and spoilage. Local foods avoid the extensive energy and pollution caused by long distance transportation.

Naturally clean and biodiverse foods enhance food security and quality while providing nutrient diversity to consumers. Nature has used the natural biodiversity strategy for eons to assure sustained plant production. A biodiverse food system creates robust production than can withstand a wide spectrum of stressors or invaders.

Monocultures put the entire food supply at risk. A single pest such as a fungus, mold, mildew, insect, worm or weevil can devastate a monoculture in a single growing season. A natural biodiversity strategy gives growers the freedom to choose desirable crops and avoid transgenic seeds.

Cleans and repairs ecosystems. Soil pollution costs U.S. citizens over $45 billion annually – about 50% more than the U.S. spends yearly on health research.[85] Yale professor Robert Mendelsohn and Nicholas Muller estimate air pollution (from all sources in the U.S.) damages range from $75 - $280 billion annually.[86]

Farmers do not pay for their environmental destruction costs. Current social policy imposes no taxes on extraction, waste or pollution. Society pays, or more precisely, our children will bear these costs. The current food system will leave our legacy – severely degraded ecosystems barren of valuable natural resources. The amber waves of grain on our plains will be gone, replaced by dry and dusty deserts.

An unfortunate unintended consequence of fossil foods has been the public and environmental health impacts from soil erosion. High commodity prices and crop subsidies motivate farmers to plant crops on highly erodible wetlands and hillsides. Research at Iowa State University provides evidence that erosion loss in some regions occurs at 50 to 500 times faster than topsoil forms. Soil erosion occurs at levels far beyond government estimates.[87] The Iowa Daily Erosion Project reported that a single 2007 spring storm created an average loss between 20 and 40 tons per acre in 14 Iowa townships.[88] Severe storms are becoming more frequent with climate change

Fossil foods impose the high cost of waste on farmers. Typically, crops absorb less than half the fertilizer applied and rest is wasted. Similarly, crops absorb as little as 5% of agricultural poisons. A single acre of a row crop such as corn erodes about six tons of soil each year. Industrial farming methods inflict substantial erosion and pollution on cropland and wetlands.

Some farmers try to practice conservation agriculture, but industrial farming methods undermine their efforts. Crops continually extract macro and micronutrients, while cultivation disrupts and compacts the soil, amplifying erosion. Chemical fertilizers and poisons are toxic to soil microbes, which reduces soil fertility and further degrades soil structure. When agricultural production has depleted soil nutrients and degraded soil structure, farmers must abandon the cropland.

A food production system that cleans air and water will benefit everyone. A process that regenerates degraded soil will provide many advantages.

Climate independence. Fossil food crops are not climate tolerant. Farmers have bred food crops for 11,000 years to produce within a narrow range of temperature and environmental conditions. Small changes from normal conditions seriously affect crop yields. A 2.5° F temperature increase may cause a 10% crop loss. A 7° F increase can cause crop failure.

Climate chaos causes crop failure from fierce storms, temperature spikes, prolonged heat, drought, hot dry winds, irrigation salt invasion, rising oceans and wildfires. Fossil foods fail with climate change, leading to hunger and malnutrition.

The only way to assure food security with global climate chaos is to free farmers with climate independence. The food production system should produce reliably independent of climate, weather, geography, or politics.

The next chapter explores how and why freedom foods support the Food Bill of Rights.

Chapter 3. Freedom Foods with Abundance

How can we transform our food supply to healthy, sustainable and food secure?

The Freedom Foods Revolution acts to transform our fossil food production system to healthy, clean, local and regenerative. Freedom Foods enables us to leave a positive legacy for our children – healthy food, clean ecosystems and abundant natural resources.

Freedom foods growers cultivate algae and other microflora in abundance microfarms using plentiful resources that will not run out.[89] Microfarmers recycle organic inputs from farm or other waste streams that are free, low-cost, or surplus. Growers cultivate microorganisms such as algae and the microflora they attract to produce food for people, feed for fish, fowl, dairy, and meat animals. Other growers grow and flow their culture to produce rich organic fertilizer for gardens or fields.

Microfarms scale to any size any size and fit in nearly any space with access to sunshine. Growers may produce year-round, independent of local weather or global climate chaos. Freedom Foods do not compete with industrial agriculture because growers use non-fossil inputs, where possible.

When abundance microfarms are developed, they will give growers the freedom to produce a natural diversity of food, feed and other sources of energy. Grower knowledge distributed globally through social networks will enable people to produce for their family and community locally.

Growers who wish to avoid eating algae directly may grow and flow the algae to feed fish or other animals. Algae feed supplements provide solutions for a litany of challenges – including enhancing animal health and vitality. Growing freedom foods for fish, fowl or meat animals avoids the harvest step. Growers simply concentrate the algae for the animals to drink. Animals are attracted to algae for to its sweet, soft, and natural taste.

Algae can provide significant improvements in yields and quality for hydroponics. Farmers growing field crops can use smartcultures to carry algae nutrients directly to the roots of their crops.[90] Algae provide a nutrient delivery system that is immediately bioavailable to the plants. Algae's full spectrum of essential nutrients assures crops avoid suffering from the problem of hidden hunger.

The excellent seed catalogue, *Cook's Garden* sells a seaweed product called Sea Magic.[91] The algae fertilizer contains cytokinins and 17 amino acids that encourage stronger, lusher growth, more sugar production in fruiting crops, and increased blooms. The company claims this seaweed fertilizer increases yields, performance and flavor. Our field and garden research supports their claims, which include 24% more tomatoes, 25% more grapes, 34% more cucumbers, and 47% more peppers.[92] Our peppers were closer to 25% yield improvement, but our garden already benefits from excellent organic soil with microorganism communities. Most nurseries and seed catalogues sell multiple algae fertilizer products made from mined fossilized seaweed. Similar seaweed fertilizers are plentiful in hydroponic supply stores and several provide similar yield boost.

Fish emulsions are also popular with gardeners but the fertilizer stinks. Why put up with the intense smell of dead and decaying fish when algae provide the original nutrients without the fishy smell?

Nutrients in fish emulsion come from algae because algae typically provide the primary nutrition for fish.

Freedom foods offer innovative solutions that support the Food Bill of Rights, which are not available with industrial foods.

Food security – Food democracy. We envision a world of abundance, with food democracy, where all people may access or grow tasty, low fat and healthy, fossil-free foods while repairing our ecosystems. We know the best foods for our families, our animals and our ecosystems are naturally biodiverse and grow low on the food chain.

Abundance production mimics nature's sustained food production for 3.7 billion years, as it recycles and reuses waste stream nutrients. Sustainable production does not help consumers unless the food or the inputs to produce food are affordable. The use of waste stream nutrients makes the inputs to produce food affordable to nearly everyone.

A distributed freedom foods production model enables food democracy and engages people globally who can produce locally. Growers use minimal fossil resources that are often surplus, affordable, and will not run out – sunshine, CO_2 and waste, brine or ocean water.

Most growing sites have sufficient sunshine for photosynthesis and those that do not can use grow lights. Modern energy production systems and manufacturing plants produce far more surplus CO_2 than growers can use. Of course, recovering the CO_2 and delivering it to distributed growers remains a considerable challenge. Growers that lack access to a waste CO_2 source may use pyrolysis where they burn organic waste in a closed kiln to produce CO_2 for algae production and hydrogen for energy.[93] Other growers may burn botanical wastes or other fuels to produce CO_2 for their culture.

The secret to affordable inputs is the recycling of waste stream nutrients. Manure typically contains about 60% of the original energy in the plant and often over 80% of the original nutrients. Growers can use solar heaters and other low cost and low energy methods for removing parasites and pathogens.

Most communities have surplus wastewater, which is rich in nutrients that algae can recover naturally. Algae absorb the organic waste and transform them into rich hydrocarbons. Growers harvest the algae, use the biomass for food and other coproducts, and can use the clean water for their households, animals or crops. In some settings such as rural Africa, the value of clean water may be higher than the green biomass produced.

Half the water stored on the planet is nutrient-rich brine water – far more than freedom food growers can use in several centuries. Salt water from estuaries and oceans offer considerable nutrients but those nutrients may be in very dilute solution. Most estuaries and coasts near human populations also suffer from severe pollution.

Abundance growers close the nutrient cycle as they recover and reuse waste stream nutrients, Figure 3.1. While harvest takes about half the nutrients from the culture or field, 95% of those nutrients are recoverable from municipal and industrial waste streams.[94] Waste stream recovery solves three major challenges: fossil resource over-consumption, nutrient costs, and pollution.

Figure 3.1 Closed Nutrient Cycle – Abundance

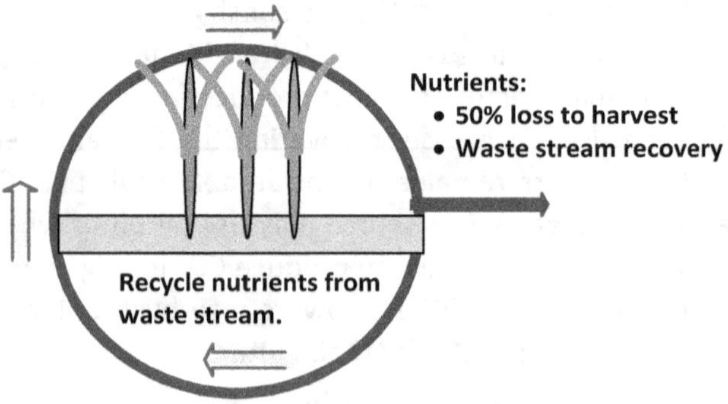

Nutrients:
- **50% loss to harvest**
- **Waste stream recovery**

Recycle nutrients from waste stream.

The human labor required to grow freedom foods is more like gardening than industrial farming. Heavy lifting and heavy equipment are generally unnecessary. With remote monitoring systems, growers do not have to be highly trained to produce consistently good biomass. Therefore, people who experience food injustice today – women, elderly, disabled and certain ethnicities – are able to grow food for their family and community.

Some unanticipated issues are certain to arise in abundance production but growers avoid most of the physical labor and health risks common in traditional food and energy production.

Healthy choices. Algae are tiny organisms, about five μ (microns) small. The period at the end of this sentence is about 20 μ. Algae reside at the bottom of the food web and provide food daily for 100 times more consumers than any other plant on Earth. Each alga cell grows and reproduces independently, creating significantly higher nutrient density than terrestrial crops.

Land plants must survive in challenging environments where most the cells are specialized for structure and lack nutrient density necessary for food. Terrestrial plants must invest 90% of their energy in nonfood producing components including roots, stock, stems, leaves and seed coverings. Typically, only the seed (or fruit) offers food value.

A three-meter corn stalk, including roots may weigh a kilogram. This huge biomass provides only 3 grams of protein stored as a fraction of the kernels on the cob. Most of the cellulosic material land plants use for structure goes to waste in terms of food.

Algae biomass has no need for cellulosic structure since it grows in aquatic environments, where water supports its structure. A kilogram of algae may contain 60% protein, which yields 600 grams of protein.

Nutrient dilution occurs in industrial foods because crop production continually extracts soil micronutrients that are not replaced. Crops grown in soil depleted of micronutrients cannot synthesize missing micronutrients, e.g. zinc, calcium or manganese, or the essential vitamins, minerals or trace elements. The resulting produce may have a normal appearance, but lack nutrient density, creating hidden hunger. Industrial farmers grow GE monocultures that often lack both nutritional density and nutritional diversity.

Algae grow to the limit of nutrients available. When the culture runs out of the first essential nutrient; algae stop growing and reproducing. Algae bioaccumulate nutrients at 1,000 times ambient levels, which means algae flourish in water with very dilute nutrients. Microfarmers monitor their cultures regularly to ensure plenty of micronutrients are present to support algae growth and development with no hidden hunger in the biomass.

Each alga cell packs a full set of essential nutrients. This dense nutrient package varies among the millions of algae species. Therefore, growers may select specific cultivars from among the many naturally biodiverse species that maximize the desired nutrients.

Freedom food growers produce delicious, high-protein foods with all the essential nutrients, vitamins, minerals, trace elements and antioxidants needed for health and vitality.[95] Freedom foods improve eyesight, respiratory capacity, cardiovascular health and brain function.[96] Freedom foods moderate or eliminate nutritional deficiencies that cause growth and development disorders, obesity, diabetes, blindness, stunting, mental retardation, and chronic fatigue.[97]

Algae contain more beta-carotene (provitamin A) than other foods. Doctors and dietitians recommend algae powder as a source of beta-carotene in dietary supplements and functional foods. Algae are rich in antioxidant vitamins (C and E), in concentrations higher than land plants. Vitamin C helps people avoid scurvy. Vitamin E moderates neurological problems due to poor nerve conduction and anemia due to oxidative damage to red blood cells.

Algae are a good source of all seven B vitamins, including vitamin B12. Algae are unique as a plant source of vitamin B12. Doctors and dieticians recommend algae, particularly nori, as a dietary supplement for vegetarians who desire to obtain vitamin B12 from a natural, non-animal source.

Algae provide a mineral profile superior to that of land plants and even milk or soybeans. Minerals in terrestrial foods such as food grains have minerals such as iron bound up in phytic acid complexes, limiting their bioavailability. These complexes cannot be absorbed into the blood stream and pass through the body. Studies show iron absorption significantly higher for marine algae compared to rice.

Algae are also rich in iodine and selenium, critical trace elements that are highly variable in food supplies by geographic region. These minerals have been associated with endemic deficiency disorders throughout history. Algae concentrate these trace minerals and only small amounts of algae (one tablespoon of dried algae) provide sufficient levels of these nutrients when introduced into the diet. Consumers may extract vitamins and nutrients from non-digestible algae by chewing algae in a cud similar to chewing gum.

Algae have a high content of glutamic acid that stimulates taste receptors with umami, (savory or hearty), amplifying taste differentiation and the desire to consume algae for its good taste.[98]

Spiritual and dietary choice. Freedom foods liberate consumers to make food choices aligned with their spirituality and dietary values. Consumers may choose from a wide variety of foods that are vegetarian, vegan, Kosher, organic or other. Each tiny alga cell concentrates the full range of essential nutrients for plants, animals and people.

Excellent sensory appeal. Food scientists attribute the lack of color, aroma, taste and texture to micronutrient deficiencies in the soil. Pictures of field tomatoes 30 years ago show tomatoes with beautiful color in contrast to the pale modern tomatoes. Consumers report that tomatoes grown 30 years ago looked and tasted much better than modern tomatoes. Organic growers typically provide the full range of micronutrients to their produce from composted organic material. Consequently, organic produce typically offer superior aroma, color, flavor and texture. Freedom farmers insure their cultures have access to the vital micronutrients that maximize growth and development and sensory appeal.

Microfarmers grow wide variety of micro and macroalgae, (seaweeds and sea vegetables) with excellent color, aroma, taste, and texture. Some microfarms grow algae to feed finfish or shellfish. Others grow algae as feed for fowl, dairy or meat animals. Algae provide their superior natural nutrient package, which is immediately bioavailable to the animal because algae cells are so tiny. The animals respond by growing faster and healthier and display higher sensory appeal.

Fish grown on a diet of food grains produce no omega-3 fatty acids because fish and other animals do not synthesize omega-3s. Fish fed algae concentrate the omega-3s synthesized in the algae into what consumers call fish oil. Fish oil comes from the algae oil consumed by fish. Many fish oil producers harvest the omega-3s directly from algae. The algae harvest is faster and cheaper than fish and unlike fish, does not accumulate mercury. The primary benefit from direct algae harvest is that consumers do not burp a disgusting fishy taste.

Other microfarmers recycle the farm waste stream, and grow and flow the algae to field crops. Algae act as a delivery system for crops supplying nutrients that are immediately bioavailable. Smartcultures, Sustainable MicroAlgae Regenerative Technologies, can increase crop yields and nutrient density while significantly improving the sensory attributes of produce. Smartcultures reduce the need for some fertilizers by 50% or more, reduce the need for cultivation and diminish ecological pollution by 95%.[99]

Local jobs. Distributed production promises to redistribute the wealth that is currently concentrated in large agribusinesses. Large agribusiness can still produce food, but many family or cooperative microfarms located near concentrations of consumers will also provide food, feed and other coproducts. The freedom foods industry and microfarms will spawn millions of new jobs that cannot be outsourced. These jobs need to stay close to consumers to produce high quality fresh foods and save transportation costs and energy.

People without affordable inputs to grow crops are completely dependent on others for their food. Food dependency and insecurity leads to severe hunger and malnutrition. The abundance model solves the challenge of affordable inputs to produce food. Practically every country and community has surplus wastewater, CO_2 and sunshine. Growers can produce good foods locally with these inputs.

These new green jobs will provide a living wage because the distributed food production model rewards local growers. The model gives freedom physical risk because growers avoid heavy equipment, huge engines, physical labor and fatigue. Growers receive no exposure to agricultural chemicals, particulates and poisons since they are not necessary. Abundance methods practically eliminate economic risk from crop failure because growers may harvest half the culture's biomass daily, year round. An occasional culture contamination or vacation may halt production temporarily but it can begin again in a matter of days.

The United Nations Millennium Project suggests that in order to resolve hunger and malnutrition, wealthy nations buy the inputs to produce food for needy nations.[100] Giving food inputs represents a far superior plan than gifting food, as the U.S. does currently with heavily subsidized crops. Unfortunately, most nations have neither sufficient wealth nor the fossil resources to provide enough inputs to produce food. As fossil resources become scarcer and more expensive, the possibility of giving food inputs becomes increasingly unattainable.

Abundance methods that produce freedom foods offer a novel solution to poverty, hunger and malnutrition. The freedom foods model holds promise for sustainable jobs and food democracy.

Preserves natural resources. Freedom foods minimize the use of fossil resources and save those precious resources for our children. Growers avoid consuming constantly more expensive natural resources and do not contribute to resource scarcity because they recycle energy and nutrients.

Fossil-free foods offer many advantages besides resource conservation. Freedom foods produced with abundance methods avoid fossil resource consumption, waste and pollution. Fossil-free foods are healthier for growers and consumers because they do not use heavy applications of freshwater, fossil fuels, chemical fertilizers and poisons. Recycling resources minimizes or eliminates the massive nutrient and chemical waste and pollution caused by fossil foods. Growers save significant costs by recycling resources instead of buying a new set with every crop.

Abundance methods enable growers to produce good foods while cleaning air and water. Each ton of algae sequesters two tons of CO_2, plus other greenhouse gasses such as methane. When consumers use the algae biomass for food, feed, fertilizer or energy, the CO_2 is actually recycled rather than sequestered. Ecologists note that every pound of algae displaces a pound of fossil foods or fossil fuels – which diminishes the carbon and ecological footprint.

Algae can remediate wastewater using no or minimal fossil energy. When producers harvest the algae, the water is clean enough for household, yard or garden use or field crop irrigation. Smartcultures enable growers to repair and regenerate degraded soils and ecosystems.

Freedom foods enable growers and consumers to eat healthfully and heartily, yet leave the ecological footprint of a butterfly.

Fresh and local. Covered or closed microfarms enable growers to produce nutrient-rich biomass in nearly any geography. Algae grow all over the earth. Kelp forests flourish under the North polar cap and other algae provide food for krill and many other creatures in cold environments. Algae crusts and liken, (an algae-fungi symbiosis) thrive in the hottest deserts.

The manna from heaven described in the Bible probably came from algae crusts and lichen. These organisms could have used the early morning mist for water and extracted nutrients from the desert soil. The protein and rich nutrients could have sustained the Israelites in the desert.

Asian societies eat more than 100 varieties of delicious fresh algae called sea vegetables. Farmers harvest sea vegetables locally from natural stands. Some seaweed, such as Nori is cultivated on nets anchored in estuaries. Algae dry quickly in the sun without losing their nutrients and without preservatives. Algae reconstitute easily in water and typically return to their fresh color.

Algae's popularity has overtaxed natural stands. Increasingly, farmers are finding ways to cultivate algae in order to meet higher demand.

Most current production occurs in estuaries and along coastlines in seawater. Microfarms offer an alternative for nearly any geography.

Local production solves three challenges: avoidance of packaging, the need to use preservatives, and reduces transportation costs substantially.

Our freedom foods goal is to enable growers to produce 50% of food within 50 miles of consumers in the U.S. Local production of half our food can save 36 billion gallons of diesel fuel a year currently used to transport foods long distances. Removing all those trucks from the road will lower road maintenance costs and truck-car crashes. Cities will benefit from fewer trucks clogging the streets. People's health will benefit from diminished smog and particulate pollution from exhaust smoke and gases.

Naturally clean and biodiverse. Microfarmers have considerable freedom of choice in species selection because scientists estimate the number of algae species at possibly 10 million.[101] Algae are extraordinarily adaptable, so growers may train a species to adapt to local growing conditions or produce more of a desirable compound such as oil or Omega-3 fatty acids.

Natural biodiversity solves four critical challenges: avoidance of monocultures and GE seeds while increasing nutrient diversity and providing clean food. Scientists find it difficult to grow algae monocultures in their laboratories because algae attract both competing species and predators. Microfarms grow biodiverse cultures because algae attract so many other organisms. These diverse microorganisms work symbiotically to support one another and their ecosystem. When the biomass is used as feed or fertilizer, the diverse organism community works together to supply essential nutrients, vitamins, minerals, antioxidants and trace elements.

Algae's natural biodiversity and adaptability enable growers to select or adapt cultures that produce the components they desire. With millions of species to choose from, growers do not have to use transgenic seeds or cultures to accomplish their production goals.

Monocultures of GE crops have produced significant nutrient dilution and diminished nutrient diversity. Microfarmers typically grow algae indigenous to their area because the plants have adapted over eons to the local microclimate. Algae-based foods provide very high nutrient density and diversity.

Naturally clean food also solves the problem of chemical fertilizer and poison residue on and in produce and refined products. Freedom foods grow with primarily natural, organic inputs.

Algae may be cultivated in dirty wastewater that contains a variety of pharmaceuticals. About 70% of all antibiotics produced in the U.S. (nearly 25 million pounds a year) are fed to cattle, pigs and poultry to boost production.[102]. Those pharmaceuticals wind up in the manure. Food crops such as corn, potatoes and lettuce absorb and concentrate antibiotics when grown in soil fertilized with livestock manure.[103]

Algae are so tiny; the individual alga cells cannot absorb a huge compound like an antibiotic. Algae absorb the nutrients by deconstructing the large compounds, one element at a time. The process detoxifies complex compounds. Local and remote monitoring systems provide an important quality assurance safeguard to ensure the foods are clean.

If the waste stream includes toxic heavy metals such as arsenic, lead, plutonium, cadmium or mercury, algae will bioaccumulate the metal. Fortunately, once the metal has moved from a dilute liquid to a solid state in the algae biomass, simple processing technology can remove the metals. Some algae producers use a business model where algae remove toxic heavy metals from industrial waste streams. The grower recovers the waste stream metals and sells the recycled metals back to the market. These microfarmers may make more revenue from the recycled metals than they do from the algae biomass.

Cleans and repairs ecosystems. The primary application for algae in the U.S. over the past 50 years has been water reclamation. Algae absorb organics in wastewater, enabling waste treatment facilities to avoid using expensive chemicals and fossil fuels. Algae waste treatment can take a cost and transform it into a profit center.

Since each ton of algae sequesters two tons of CO_2, algae can clean air. Several companies produce algae by flueing power plant or manufacturing smoke stacks through algae production facilities to feed algae.

Farmers that apply smartcultures methods grow algae near their fields and flow the algae along with the irrigation water.[104] Algae act as a very efficient nutrient delivery system that brings target nutrients just when the crop needs them.

Industrial agriculture degrades fields to the point of abandonment due to systemic disruption, extraction, compaction, and erosion. Fields become depleted of topsoil, micronutrients and humus (organic matter) because industrial farmers seldom replenish them. Fields deficient of humus cannot hold soil moisture and require substantially more irrigation per pound of crop. Large tractors and irrigation compacts soil, which amplifies erosion and limits root development.

Algae use a multifunction strategy to regenerate degraded soil.[105] Algae biofertilizers replace the extracted micronutrients with nutrients that immediately bioavailable to the crop. Algae continue to grow in the field, as long as soil moisture is present, adding to soil organic matter. Algae produce polysaccharide sheaths that loosen the soil and have increased soil porosity, (looseness) by 500%. Algae also attract a broad spectrum of other microorganisms that work symbiotically to repair degraded soil.

Geography and climate independence. Closed or covered CAPS can produce food nearly anywhere. Growers have the substantial advantage of local production on non-cropland. Growers can site microfarms on rooftops, balconies, open lots, deserts or mountains because they do not need soil.

Modern field crops are dependent on a narrow temperature range to grow productively. Prolonged heat, temperature spikes or storms can destroy fossil crops. Abundance production methods are climate independent and most growers can produce year round – when the sun shines. Growers in the high latitudes can produce in greenhouses and use freedom energy LED grow lights.

Terrestrial plants cannot grow where sea or irrigation salt has invaded cropland due to a plumbing problem. Large salt ions clog up their roots, starving the plant of nutrients. Algae flourish in salt because they evolved in ancient oceans that were highly saline and algae have no roots. Algae grow in practically any kind of water.

All these benefits will evaporate unless people adopt freedom foods.

Freedom foods adoption

Many people remain skeptical that consumers will accept freedom foods, despite their many advantages. The science of consumer behavior offers only two strategies to change people's food consumption patterns:

1. **Push** people to eat lower on the food chain with influence strategies, such as carefully explaining the health and environmental benefits.

2. **Pull** people down the food chain by providing nutritious and delicious foods that appear similar, yet taste better than their equivalent fossil food.

The push strategy called cognitive conditioning, works only for a tiny percentage of consumers. Most people try diets or practice good food habits for a brief period, then become recidivists. They revert to their favorite comfort foods. The statistics on diets confirm that most do not work.[106] Food behaviors are central to family, culture and society, and are extremely difficult to change.

Push strategies have failed scientists and practitioners repeatedly. Many fat people are experts on nutrition. Repeated studies show that knowledge alone fails to change behavior. People change their behavior when they find something better as a substitute.

Push strategies cannot work because fossil foods monopolize our fields, stores, refrigerators and plates today. The only choice most consumers can make currently is industrial versus organic foods or meat versus vegetables. Industrial foods control 97% of the U.S. market and 94% in Europe. Unlike the U.S., the European Union offers organic growers subsidies for the social, environmental and health benefits provided by organic foods. Freedom foods offer a fossil free alternative, aligned with organic methods, and will give consumers another choice – once they are available.

What's it all about Algae?

The Freedom Foods Revolution proposes a pull strategy where food low on the food chain goes into products that appear similar to traditional foods but with microorganism ingredients that are natural, healthier and provide superior sensory pleasures. To gain widespread consumer adoption, freedom foods must offer more than health and environmental benefits. Consumers must perceive them as offering a bundle of positive attributes. Each additional positive characteristic will accelerate consumer adoption.

Freedom foods give consumers a choice between two chips that look identical, and offer equal sensory appeal.

Industrial Food and Freedom Food snack Chip
Which chip would you choose?

Nutritional value	Industrial food Corn chip	Freedom food Algae chip
High saturated fats	Yes	No, 85% less
High cholesterol	Yes	No, 80% less
Loaded with empty calories	Yes	No, 80% fewer
High protein	No	Yes, 50% higher
High nutrients	No	Yes, 100% higher
Omega-3s	No	Yes, 100% higher
Vitamins and minerals	No	Yes, 100% higher
Trace elements	No	Yes, 100% higher
Diminishes food cravings	No, increases	Yes, 100% lower

Unlike the corn chip, which comes from a monoculture, the algae chip comes from a healthier, naturally diverse culture. The algae chip has no genetically engineered material and no agricultural chemical or pesticide residue. The algae chip growth and production consumed minimal fossil resources and no pollution. The growing process for the algae chip improved the quality of the surrounding air, water, and soil.

When consumers have the freedom to choose foods with superior sensory appeal and better nutrition at a reasonable cost, they are likely to choose freedom foods.

Freedom foods will attract some consumers for the health benefits from eating low on the food chain. Others will choose freedom foods to avoid GE crops, preservatives and chemical residues. Some will want freedom foods to lighten their ecological footprint. The majority will probably make their choice based on superior nutrition and taste.

The next chapter explores algae, the most abundant and nutritious plant on our planet – algae.

Chapter 4. What's it all about Algae?

Algae, nature's oldest and possibly best food.

The foundation for freedom foods and abundance growing methods relies on a plant that used 3.7 billion years of evolution wisely to develop strategies to:

- Adapt to acute searing and freezing temperature spikes.
- Survive the extremely hot temperatures of early Earth.
- Grow in ocean, brine, saline, waste and fresh water.
- Live through brutal electrical, wind and ice storms.
- Use either organic material or sunshine for energy
- Go dormant when conditions degrade, yet survive.
- Grow rapidly at any latitude, longitude or altitude.
- Grow faster than any other plant on Earth.
- Thrive using no extracted fossil resources.
- Maximize productivity per unit of space.

All living things evolved from algae so it should be no surprise that algae contain all the essential nutrients for life and vitality. A few grams of algae a day act as a nutritional supplement, providing the essential nutrients, vitamins, minerals and antioxidants.

Each day, algae create 70% of the world's oxygen, more than all the forests and fields combined. Algae synthesize roughly 0.8×10^{11} tons of organic matter daily, constituting about 40% of the total fresh organic matter grown on our planet.[107]

Microalgae

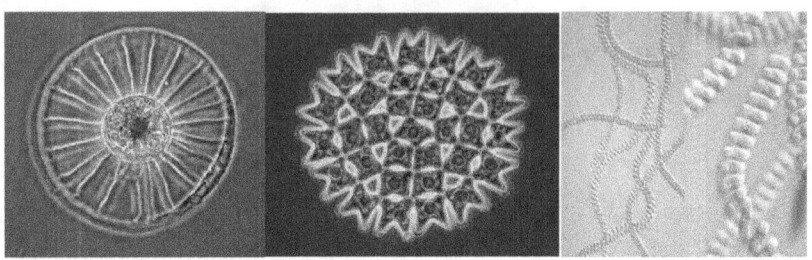

The first algal (singular) cell was among the earliest life forms on Earth, probably about 3.7 billion years ago in oceanic environment synthesized by abiotic, high-energy processes including lightning, ultraviolet radiation and pressure shock. The atmosphere was anaerobic with high levels of methane, hydrogen and ammonia but no oxygen.

Algae, often called microscopic phytoplankton, grow in most bodies of water, moist places, on and in trees, and even in rocks. This little plant provides the foundation for the food chain, feeding both microbial and animal plankton; zooplankton and fish. Subtract algae and phytoplankton from the water column and fish, shellfish, reptiles and other aquatic creatures cannot survive.

Algae provide a full spectrum of protein, nutrients, vitamins and minerals because trillions of hungry consumers depend on algae to sustain their vitality. Over 100 times more creatures eat algae or algae feeders than any other food. Algae deliver protein energy and essential nutrients to the smallest krill, as well as to the largest animal on our planet – the great blue whale.

Algae grow all over the Earth in marine or fresh water habitats or on land, when moisture is available. Unlike land plants that die without water, algae simply go dormant, wait for moisture, and then begin their rapid development. Algae grow to the limit of the nutrient supply, and then pause or go dormant until conditions improve.

Since algae form the bottom of the food chain, everything around acts as a predator. Algae's strategy to predation is brilliant – grow faster than consumers can eat. A single alga cell may produce one million offspring in a day. At night, algae take a well-deserved rest in a phase called respiration. While an individual alga cell is not visible, algae communities appear first as a cloud and then as tiny specks that are cell clusters. Some algae aggregate to form structures, such as filaments, globes, wheels or with spirulina, spirals.

Algae break the rules for plant classification because they evolved in many different forms – cells, multicellular plants, bacteria and in nearly infinite combinations. While the various species share certain characteristics, different algae display extraordinary variety in shape, size, structure, composition and color.

Many species are single-celled and microscopic including phytoplankton and other microalgae while others are multicellular and may grow large such as kelp and Sargassum. Phycology, the study of algae, includes the study of prokaryotic forms known as blue-green algae or cyanobacteria. Some algae also live in symbiosis with lichens, corals and sponges. The basic single-celled organism, algae, has the general appearance illustrated in Figure 4.1. The University of Montreal, U.C. Berkeley, University of Texas and others host culture collections of algal species with descriptive details and pictures.[108]

Figure 4.1 Algae Cell

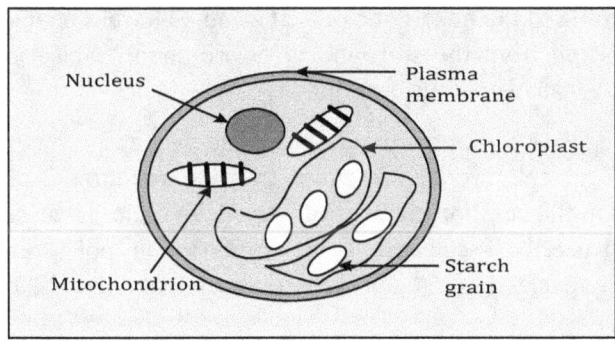

Eukaryotic green algae (Greek for "true nut") plants have cells with their genetic material organized in organelles. They create discrete structures with specific functions and have a double membrane-bound nucleus or nuclei. The prokaryotic cells of blue-green algae, cyanobacteria, contain no nucleus or other membrane-bound organelles.[109] Algae can be lively little critters even though they are not animals. Many can swim, such as dinoflagellates that have little whip-like structures called flagella. Some use the flagella to pull or push themselves through the water. Some algae squish part of their body forwards and crawl along solid surfaces.

Other species are made of fine filaments with cells joined from end to end. Some clump together to form colonies while others float independently. Seaweeds may grow in nearly any shape such as cones, tubes, filaments, circles or may imitate the shape of land plants. Seaweeds developed in parallel evolution with land plants.

Algal Cell Walls

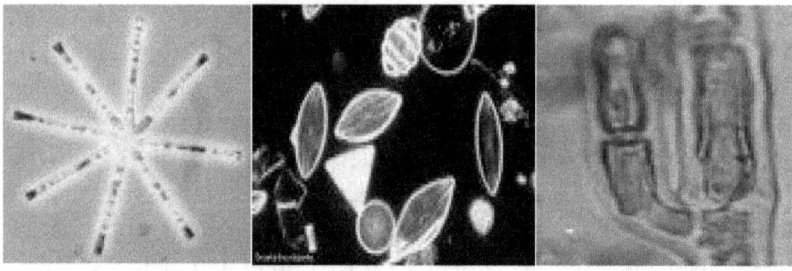

Major steps in cell complexity occurred with the evolutionary progression from a virus to bacterium and then from the prokaryotic cells of bacteria to the eukaryotic cells of algae. Cell walls enable algae to protect itself from the surrounding environment, typically water and pressure, called osmotic pressure.

Cell walls regulate osmotic pressure produced by water trying to flow in or out of the cell through its semi-permeable membranes due to a differential in the solution concentrations. Algae typically possess cell walls constructed of cellulose, glycoproteins and polysaccharides while some species have a cell wall composed of silicic (silicon) or alginic acid.

Classifications

The major groups of algae have been distinguished traditionally based on pigmentation, shape, structure, cell wall composition, flagellar characteristics, and products stored. Algae display so many variations, even within each species, that they express exceptions to nearly every classification rule.

Red algae, for example, are a large group of about 10,000 species of mostly multicellular, marine algae, including seaweed. These include coralline algae, which live symbiotically with corals, secrete calcium carbonate and play a major role in building coral reefs. Red algae such as dulse *(Palmaria palmata)* and laver (nori/gim) are a traditional part of European and Asian cuisine and are used to make other products such as agar, carrageenans and other food additives.

The broad algae classification includes:

- Bacillariophyta – diatoms
- Charophyta – stoneworts
- Chlorophyta – green algae
- Chrysophyta – golden algae
- Cyanobacteria – blue-green
- Dinophyta – dinoflagellates
- Phaeophyta – brown algae
- Rhodophyta – red algae

Diatoms, stoneworts and dinoflagellates

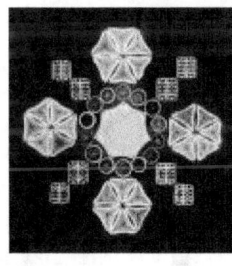  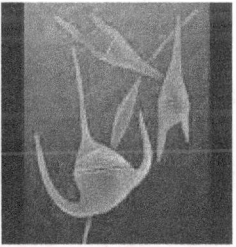

Productivity

Scientists have known algae's food value for centuries and food potential for at least 100 years. Consider the annual protein production per acre for food grains calculated using half its theoretical photosynthetic capacity, Figure 4.2. Algae provide a superior set of vitamins and minerals than found in land plants. Algae are not a full solution for malnutrition because the biomass is freedom on calories. Fortunately, calories are cheap and easy to add to a diet.

Figure 4.2 Algae Protein Production Potential –
Pounds per Acre per Year

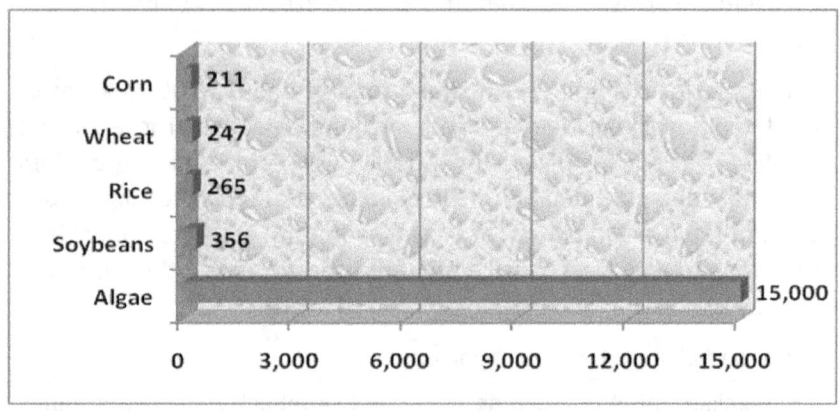

Algae flours are extremely malleable in the sense that algae can substitute for wheat, corn, rice or soy products while providing higher protein and a higher quality nutrient profile. Algae foods may include protein-rich milk, ice cream, chocolate (with superb taste and 80% less fat), baked goods of any size, shape or texture such as tortillas, crackers or cakes. The biomass may provide texturized vegetable protein with added fiber or extruded to make additives for meats that improve moisture retention and increase protein while lowering fat and cholesterol.

Processing algae can match the form of nearly any food such as peanuts, pasta, pesto or protein bars. Fortunately, years of food processing experience with terrestrial crops that have an unappealing natural taste, such as soybeans, make it easy to add colors, flavors, textures (fibers) and aromas.

Land plants evolved from algae about 500 million years ago.[110] Land plants have specialized cells for moving nutrients and for reproduction that algae do not need. Algae are distinguished from the higher plants by a lack of true roots, stems, or leaves. Some seaweed appear to have leaves or trunk but they are pseudo leaves made up of the same cellular structure as the rest of the plant.

Algae use nitrogen to manufacture amino acids, nucleic acids, chlorophyll and other nitrogen compounds. Cyanobacteria are able to fix nitrogen absorbed from the air, as well as from water, in a process known as diazotrophy. Since the atmosphere is nearly 80% nitrogen, nitrogen fixing is a strong competitive advantage for growth because water-based nitrogen is often limited.

Nitrogen fixing also means that the plant biomass has value as a low energy input, high nitrogen fertilizer because algae fixes nitrogen naturally, without added energy. About 90% of the cost of commercial synthetic fertilizers comes from the energy, typically natural gas, used to extract nitrogen from the air.

Variation

Algae range from microscopic single-celled organisms to multicelled organisms and to 60-meter kelps. These plants thrive all over the world in marine and fresh water environments – nearly any moist environment. Terrestrial algae grows in all types of soils where they can capture nitrogen from the air that can be used through the roots of plants. They may be free-living or live in symbiotic association with a variety of other organisms such as lichens and corals.

Algal Shapes

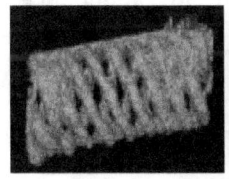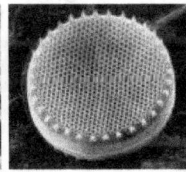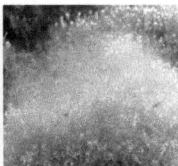

Each species may exhibit multiple strains with unique characteristics. A single strain may display completely different structural expression and composition in different growing conditions with variations in light, temperature, nutrients, mixing or water pH.[111]

Macroalgae

About 10% of algae species are macroalgae, (seaweed and sea vegetables) such as kelp that can grow to 60 meters. Most are microalgae that occur in every color, shape and small size imaginable. Each of the estimated 10 million algae species provides its unique nutrient profile as well as the many compounds stored in its biomass.

Macroalgae

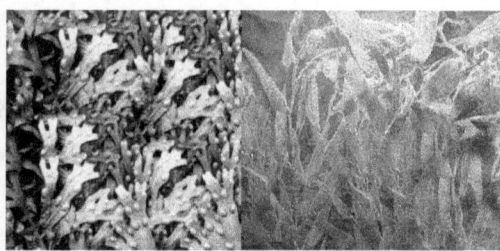

Away from the oceans, most algae grow in, on or among the roots of land plants. A handful of local dirt may hold 100 algae species and several billion alga cells. Land plants need algae to break down chemical fertilizers so they are bioavailable and absorbable by the plant. Land plants and algae work symbiotically as algae supplies nutrients and plants provide a protected area in which to grow. Algae support symbiotic relationships with mosses, fungi, yeasts, lichen, corals and sponges.

The next chapter introduces algae cultivation in abundance microfarms.

Chapter 5. What are Abundance Microfarms?

Envision 10 million Green Masterminds growing food and other valuable coproducts in abundance microfarms globally. Imagine what each can do for the health and vitality of their family and community locally.

Abundance growers mimic nature and use the oldest, simplest, yet most productive growing system on Earth – photosynthetic microorganisms. Growers practicing abundance are essentially green solar gardeners as they transform solar energy to rich, nutritious plant biomass. The green biomass concentrates energy in chemical bonds that are portable and may be used directly for food or transformed to many other forms of energy.

Microfarms cultivate communities of microorganisms similar to the algae that feed plants in the fields and the microflora that provide us nutrients in our gut. Microfarmers train indigenous, local algae to produce proteins, oils, carbohydrates and other coproducts rapidly. Microfarms grow food, nutrients, feed, fodder, fertilizer, biofuels, nutraceuticals, medicines and advanced compounds, Figure 5.1.

Microfarms scale to nearly any location and may be small, medium, or enormous growing systems that serve a family, village, community or city. The footprint may fit in a corner of a backyard, rooftop, balcony, barn, field, wetland, desert, prairie or other non-crop land.

This chapter is adapted from *Abundance Microfarms* by Mark Edwards and Robert Henrikson. Microfarms are not available yet but we are working on the technology at www.AlgaeCompetition.com.

Figure 5.1 Abundance Microfarm

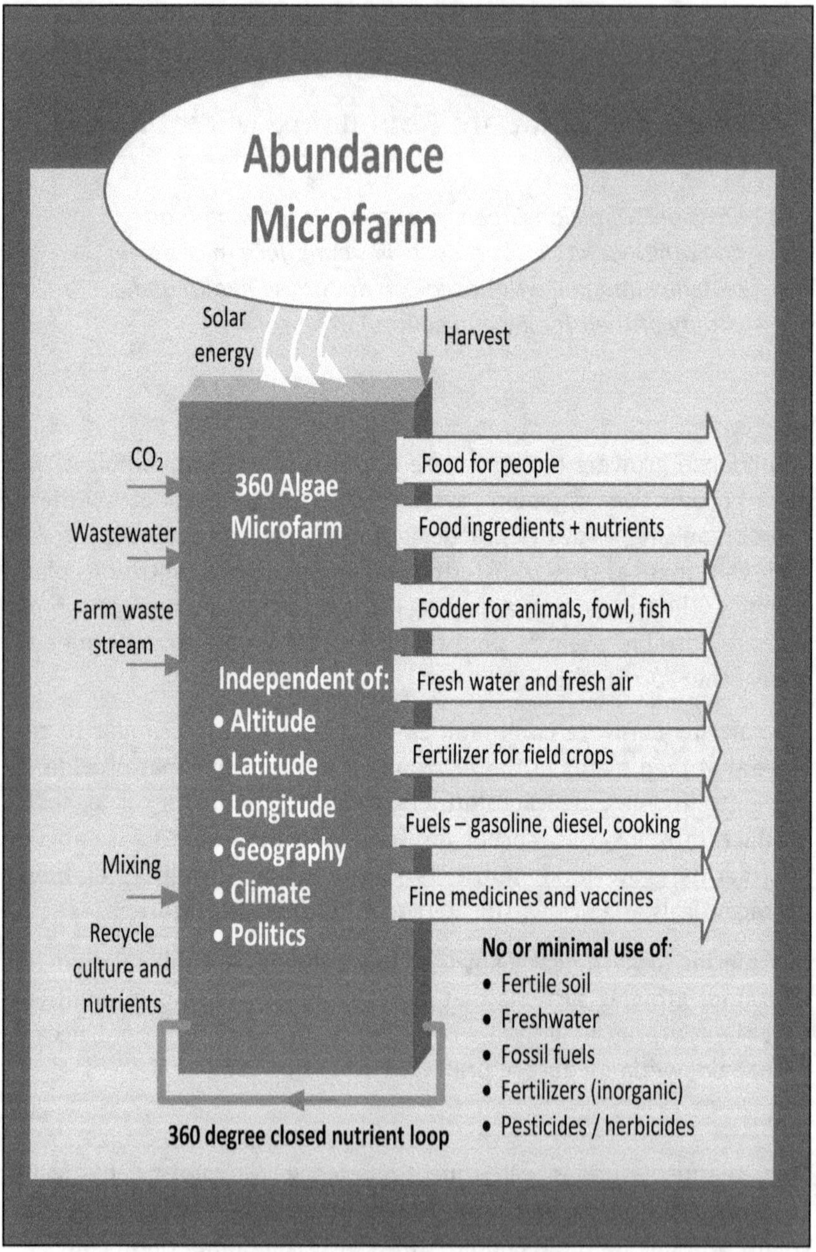

Abundance Microfarm

Solar energy

Harvest

CO₂

360 Algae Microfarm

Wastewater

Farm waste stream

Independent of:
- Altitude
- Latitude
- Longitude
- Geography
- Climate
- Politics

Mixing

Recycle culture and nutrients

Food for people

Food ingredients + nutrients

Fodder for animals, fowl, fish

Fresh water and fresh air

Fertilizer for field crops

Fuels – gasoline, diesel, cooking

Fine medicines and vaccines

No or minimal use of:
- Fertile soil
- Freshwater
- Fossil fuels
- Fertilizers (inorganic)
- Pesticides / herbicides

360 degree closed nutrient loop

Algae energy

Abundance cultivates the fastest-growing plant on the planet to provide portable energy usable in a multitude of ways, including:

- **People** – organic protein, nutrients and micronutrients in food.
- **Animals** – organic protein and nutrients in fodder.
- **Fowl** – natural protein and nutrients for birds.
- **Fish** – natural protein and nutrients in fish feed.
- **Land plants** – rich, full spectrum organic fertilizer.
- **Fire** – high-energy algae oil for cooking and heating.
- **Cars** – lipids and carbohydrates refined to biofuels.
- **Trucks and tractors** – high-energy clean, green diesel.
- **Trains, boats and ships** – high-energy clean diesel.
- **Planes** – high-energy, clean aviation gas and jet fuel.

Any product made from fossil fuels can be made from algae because nature chose algae as the primary feedstock for fossil fuels. Commercial producers are excited about replacing fossil fuels with algae. However, human societies survived for many millennia without the convenience energy sources derived from fossil fuels. The most critical energy source for humans is food. We survive only a short time when deprived of the vital energy supplied by food.

Microfarmers use cultivated abundance production systems, (CAPS) to grow algae. Growers cultivate microbial communities, which may be pure strains of algae, but are often diverse communities of algae and the multitude other microorganisms algae attract. Microflora communities thrive in aquatic and moist terrestrial settings and include algae, fungi, bacteria, viruses, slimes and other tiny organisms. This diverse array of microorganisms works symbiotically to produce compounds valuable to plants, animals, fish and humans.

Industrial or organic farmers may use abundance methods to recover and reuse the energy and nutrients in the farm waste stream to reduce production costs while improving soil fertility, crop yields and produce taste, nutrition and quality. Urban gardeners may source nutrients from municipal and industrial waste streams to grow rich algae biofertilizers that speed plant growth and development as well as increase produce size, weight, taste, texture, color, nutrition and quality.

Microfarmers cultivate algae and possibly other microorganisms as they follow one or a combination of four sustainable and affordable food and energy, (SAFE) production paths, Figure 5.2.

Figure 5.2 Abundance SAFE Production Paths

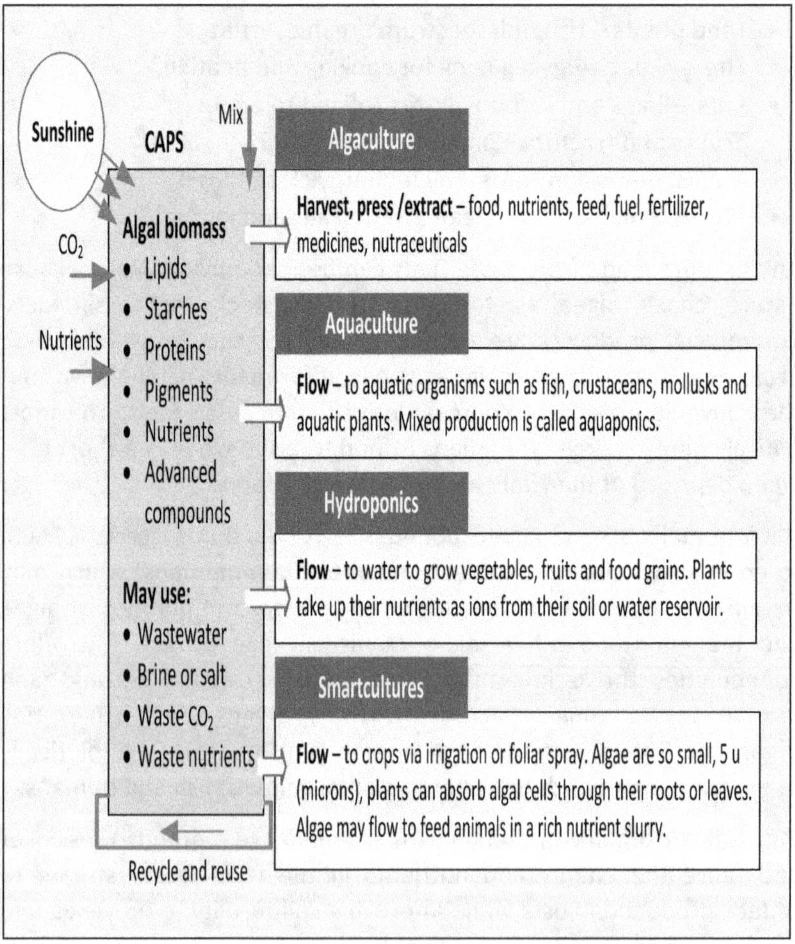

Sunshine

CAPS   Mix

CO_2

Nutrients

Algal biomass
- Lipids
- Starches
- Proteins
- Pigments
- Nutrients
- Advanced compounds

May use:
- Wastewater
- Brine or salt
- Waste CO_2
- Waste nutrients

Recycle and reuse

Algaculture

Harvest, press /extract – food, nutrients, feed, fuel, fertilizer, medicines, nutraceuticals

Aquaculture

Flow – to aquatic organisms such as fish, crustaceans, mollusks and aquatic plants. Mixed production is called aquaponics.

Hydroponics

Flow – to water to grow vegetables, fruits and food grains. Plants take up their nutrients as ions from their soil or water reservoir.

Smartcultures

Flow – to crops via irrigation or foliar spray. Algae are so small, 5 u (microns), plants can absorb algal cells through their roots or leaves. Algae may flow to feed animals in a rich nutrient slurry.

What are Abundance Microfarms?

Algaculture

Algaculture grows microalgae or macroalgae, (seaweed), for commercial purposes or domestic needs. One-third of the algae grown commercially currently goes to feed fish and shellfish. Extraction of the algae biomass enables the farmer to use the energy, nutrients and various coproducts for local needs, Figure 5.3. Algaculture producers use many CAPS shapes, sizes and forms including ponds, troughs, semi-closed and closed systems. Food, health food, feed and nutraceutical producers include Earthrise, Phyco Biosciences, Origin Oil, Aurora Algae, Solazyme, Seambiotic, Cellana, and Martek Biosciences.

Microfarms sited near a carbon source such as a waste pile on a farm or a coal fired power, cement or manufacturing plant gain the advantage of a free carbon source. Every ton of algae sequesters nearly two tons of CO_2, so free carbon reduces operational costs.

Others may site cultures near a wastewater treatment facility to gain access to free nutrients. Some growers may source organic wastes from farm or other waste streams. Algae grow well in fresh water but communities have competing needs for sweet water. Many communities have substantial sources of gray brackish or wastewater that is not potable but excellent for growing algae. Some farms have reservoirs, ponds or wetlands that capture farm runoff, which are perfect for growing algae.

Half of the water stored in the earth's crust is brine water, which is too salty for human use or for irrigation. Algae thrive on brine water, which often carries the full spectrum of essential nutrients. Many deserts, including those in the U.S. Southwest, have huge underlying oceans of brine water in relatively shallow aquifers. These brine aquifers could produce millions of tons of algae biomass.

Growing algae as fodder for animals, birds or aquatic creatures will be popular in many settings because animal fodder requires lower levels of cleanliness, (except for pets in the U.S.), than producing food for direct human consumption.

Table 5.3 Algae for Food, Biofuels and Novel Solutions

Food	Biofuels	Novel Solutions
Primary • Protein • Lipids – oils • Carbohydrates • Nucleic acids **Secondary** • Flour • Meat enhancer • Ice cream • Milk substitute • Sugar substitute • Sea vegetables • Food ingredients • Emulsifiers and thickeners • Novel flavors and textures • Pigments • Health foods • Nutraceuticals • Omega 3s **Feed and fodder** • Pets, fish, fowl • Meat animals • Micronutrients • Medicines and vaccines	**Primary** • Gasoline • Clean diesel • Methanol/ethanol • JP-8 jet fuel **Secondary** • Aviation gasoline • Alcohols • Hydrogen • Asphalt • Plastics, biodegradable • Rubber substitute **Biofertilizers** • Organic N-P-K • Bioavailable target nutrients • Micronutrients • Plant hormones • Soil organics • Build soil structure • Improve porosity • Plant growth regulators • Natural pesticides • Natural herbicides	**Air** • Carbon sequestration • Carbon capture/recycle • Capture sulfur • Capture heavy metals **Water – clean** • Waste streams – municipal, industrial, farm, brine and ocean • Recover heavy metals **Cosmetics** • Moisturizers • Skin care **Local algae production** • Foreign aid • Disaster relief • Hunger and poverty **Medicines** • HIV / AIDS and SARS • Vaccines • Antibiotics /antiviral • Burns and bruises • Stomach remedies • Anti-cancer toxins • Pharmaceuticals • Advanced compounds

Wastewater CAPS can produce food quality algae with the proper safeguard mechanisms in place. Many human-grade valuable coproducts may be extracted from wastewater algae such as vitamins, minerals, antioxidants, trace elements pigments, oil and carrageen.

Hydroponics

Algaculture producers may produce a slurry or solid product similar to fish fertilizer for use in hydroponics. Some growers grow algae next to their hydroponics unit and a portion of the algaculture flow to containers where vegetables, grains and fruits grow in the rich algae water. Algae provide all of the macro and micronutrients necessary to grow large, colorful and tasty produce. Farmers have been using smelly fish fertilizer for decades to improve plant germination, growth and yields. Algae provide a less expensive alternative with a better nutrient profile for plants than fish oil with a pleasant organic smell.

Grow algae to feed vegetables in water – hydroponics

Hydroponic farmers grow plants using mineral solutions in water rather than soil. In natural conditions, soil acts as a mineral nutrient reservoir but the soil itself is not essential for plant growth. Terrestrial plants grow well with their roots in an inert medium such as perlite, gravel, mineral wool or nutrient solution. Research shows hydroponic crop yields are be no better than crop field yields with good soils because crop yields are ultimately limited by factors other than mineral nutrients; especially light. Later research showed that hydroponics offers other advantages, including constant access to oxygen, and that the plants have access to as much or as little water and nutrients as they need. Hydroponics growers produced vegetables on Pacific volcanic islands that lacked fertile soil in World War II. Hydroponics saved considerable transportation cost.

Aeroponics, developed largely by NASA for space travel, grows plants in an air or fine mist environment without soil or aggregate medium. Aeroponics culture differs from both hydroponics and *invitro* production (plant tissue culture). Unlike hydroponics that uses water as growing medium and essential minerals to sustain plant growth, aeroponics cultures grow without an aggregate medium. Growers transmit nutrients by water mist, so aeroponics is actually a form of hydroponics.

Aquaculture and Aquaponics

Aquaculture farmers grow fish and shellfish that feed on aquatic plants such as algae. Algae represent the preferred diet for most fish fry (immature fish) because the cells are small enough for the fry to eat. Most fish evolved on an algae diet in their natural settings. Most fish grow faster and have fewer digestive problems on algae compared with food grains.

Growing algae to feed fish – Aquaculture

The Chinese have practiced aquaculture since 2500 BC. Today, half the world's commercial fish and shellfish production comes from aquaculture. A recent scientific study reported that over 90% of the large fish have been extracted from the oceans. Unfortunately, fishermen overharvest many of the smaller fish too, depleting the food chain. With diminishing natural fisheries in oceans, rivers, lakes and estuaries, aquaculture will play a larger role in our food supply.

What are Abundance Microfarms?

Aquaponics integrates fish and plant farming. Farmers grow algae to feed fish that add urea to the water. The nitrogen rich water flows to hydroponic greenhouses where vegetables and fruits grow in the high nutrient water. Polycultures can grow food with renewable energy and in closed systems, minimize consumption of fossil resources, including power and fresh water.

Smartcultures

Sustainable MicroAlgae Regenerative Technologies, (smartcultures), enable field crop farmers to recover, recycle and reuse the energy and nutrients in their farm's waste stream to improve crop quality, taste and yields while reducing operational costs by 30 to 50%.[112] Animal farmers can recover most of the energy and nutrients remaining in the farm waste stream and recycle it to feed farm, dairy, poultry or meat animals.

Smartculture farmers skip the harvest step and simply "grow and flow" the algae culture directly to their fields to recycle organic fertilizer to their crops that is immediately bioavailable to the plants. Smartcultures deliver 74 nutrients and trace elements that plants use to grow faster and stronger and produce higher yields.

Smartcultures employ a set of technologies that imitate nature to provide enhanced foundation (soil structure) and food (nutrients) to plants. Every farmer and gardener knows plants thrive in amended soils; they grow faster, stronger, and larger, and they have better taste and texture.

Smartcultures begin at the crop foundation – soil – with tiny microflora attracted by algae in plant roots that are ingeniously self-regulating and self-regenerative. Smartcultures move farmers toward abundance production by significantly reducing, but not eliminating the use of fossil resources for growing field crops. Farmers using smartcultures are able to leave every field better than they found it.

Abundance growers of field crops can use 80 to 100% fewer agricultural chemicals because algae biofertilizers provide growth hormones that make plants stronger and able to produce natural pest and disease defenses. Abundance growers can reduce soil compaction

500%, enabling significantly longer and stronger root structure. Stronger roots give plants a deeper reach for nutrients and soil moisture. Healthier plants on a stronger foundation need less water and are less vulnerable to weather, winds, weeds, disease and pests.

Smartcultures grow algae in the farm waste stream to fertilize fields

CAPS are portable to various waste sources. Our preliminary field research shows that farmers who practice Smartcultures methods may:

- Increase income from 20 to 50% by improving crop yield and quality – micronutrients, vitamins, antioxidants, color, taste, texture and shelf life.
- Lower fuel consumption by 20 to 30%.
- Decrease chemical fertilizers by 30 to 50%.
- Reduce air, soil and water pollution by 80 to 95%.

Farms without irrigation systems spray the algae solution on the fields. The algae not only provide an organic fertilizer delivery system directly to the roots of crops but algae continues to grow in the field, as long as moisture is present, regenerating soils and creating additional organic material. Algae's ability to extract in situ nutrients provides a tremendous advantage. Most farmers have available waste streams from human, animal and vegetative wastes on which algae can thrive. Rather than spending 30-40% of their production costs on fertilizer, algae may cut nutrient cost in half because algae can recover 90% of the nutrients from the farm waste stream. The typical farm waste stream contains about half of the nutrients needed for the next crop. Therefore, the net fertilizer reduction approaches 50%.

Smartcultures transform agricultural methods and rather than mining and paying high prices for chemical fertilizers and using them once, farmers can continuously recycle nutrients. Rather than systemically extracting soil nutrients and organics, farmers can cultivate algae and microflora to add nutrients and organics to their fields. Rather than using chemicals that destroy soil microbes and soil structure, smartcultures cultivate microbial communities that improve soil structure. Industrial agriculture degrades soil, promotes erosion and creates severe pollution, while smartcultures improve soil structure and reduce nutrient waste, erosion and pollution.

Recent fertilizer price escalation has pushed the cost of fertilizer to 30 to 40% of farm operating costs. Farmers face two other serious problems with industrial fertilizers: bioavailability and erosion. Chemical fertilizers must first be broken down by microorganisms, e.g. algae, in the soil before they can be absorbed by the plant. The process may take months or years, so farmers have to put on far more fertilizers than the plant needs in order to maximize growth. Much of the applied fertilizer does not reach the crop and erodes with irrigation, winds and rains. The next year, the farmer must apply even more fertilizer to achieve the same yields. This model is sustainable only as long as fertilizers are cheap and soils do not wear out.

Farmers can improve the quality and quantity of field crops because algae biofertilizers are immediately bioavailable to the plants and create almost no waste. Some algae are able to unlock nutrients in the soil, such as phosphorus, (P). Over 95% of the P in some soils is locked in large molecules that are not absorbable by plants. Algae can solubilize the P and other elements, making them bioavailable.

Smartcultures can deliver precise amounts of target nutrients carried in algae biofertilizers at specific times during a crop's growing cycle – which can maximize germination, early growth, maturation and fruiting. CAPS near fields can overload one or more nutrients and deliver them to crops exactly when needed. For example, adding more calcium when the crop is fruiting may enhance fruit size, weight, color, texture and taste.

Farmers can save money and energy by lowering their consumption of fuel due to easier cultivation. In some settings, smartcultures have improved soil porosity (loosen compacted soil) by 500%. Smartcultures reduce air, soil, and water and pollution because algae biofertilizers and plant growth hormones significantly diminish the need for agricultural chemicals.

Drip irrigation can deliver algae biofertilizers precisely to the roots, minimizing water use and the waste of nutrients. Algae continue to grow in the soil while moisture is present, which adds rich organic matter and conditions the soil, making it more erosion resistant. This model may also use no or minimal till to minimize soil disruption and provide longevity to the water-efficient drip irrigation system. In non-irrigated settings, growers may apply the algae culture with a field or aerial sprayer.

We will need every sustainable form of food production to meet our future food demand, including organic production and vertical farming. Organic production consumes more cropland, freshwater and fuel than industrial agriculture but less inorganic fertilizer and agricultural chemicals Organic growers often produce local to consumers, consume few agricultural chemicals and create substantially less ecological pollution than industrial agriculture. Unfortunately, most countries, including the U.S., have insufficient cropland to grow the biomass needed for composting organic growing systems. Vertical farms offer the promise of resource efficiency and pollution free food but the cost and concentration of production may undermine profitability and viability. Time and testing will demonstrate methods of sustainable production.

Chapter 6. Why not Freedom Foods Now?

Modern foods seem cheap,
but at what cost to our children?

The first question most people ask about freedom foods is, "Why weren't freedom foods and abundance methods invented before?" Consumer behavior, the science of why people make certain choices, provides an answer. Agribusinesses promise farmers that synthetic chemical inputs trump nature. Agribusinesses drive farm policy, which unsurprisingly subsidizes industrial agriculture research to the exclusion of natural processes.

Freedom foods represent a "natural process" invented by nature and are not patentable or listed as intellectual property. Agribusinesses that supply inputs to farmers hire legions of technologists and scientists to develop synthetic compounds. These patented compounds, synthetic poisons, and GE seeds sell to farmers for premiums that create wealth for the companies and their executives. Agribusiness advertising promotes this intellectual property and creates a belief in better living and farming through chemistry. Monsanto's Round-Up™ is among the most common words in modern farming.

Consumers are addicted to highly processed foods, with high sugar, fat, cholesterol and salt but few nutrients. Farmers are addicted to increasingly expensive genetically engineered seeds, synthetic fertilizers, and patented herbicides, pesticides and fungicides. The sad irony is that the synthetic poisons kill the beneficial microbes that nature put in the field to feed and nurture plants. Imagine, paying premiums to kill the organisms that work symbiotically with plants to provide nutrients and plant hormones for their vitality and defense.

Modern farmers have bought into chemical fertilizers because they are cheap and easy to apply. Industrial agriculture systemically extracts macro and micronutrients as well as organics from field soils. Each year, many farmers replace only the three N-P-K macronutrients (nitrogen, phosphorus and potassium). With every crop, micronutrients diminish along with soil organics, which creates nutrient dilution and hidden hunger, which diminishes color, aroma, taste, and texture in produce.

Modern farmers use large, heavy tractors that cultivate quickly but compact soil, which diminishes root growth and accelerates erosion. Farmers buy tons of chemical fertilizers, pesticides, herbicides and fungicides. Crops are developing resistance to chemical fertilizers, so farmers must apply more. Pests and weeds are developing resistance to chemical poisons, which means farmers must use more or change poisons.[113] Plants often absorb less than 5% of the agricultural poisons applied to fields, which creates enormous waste and cost. The residual fertilizers and poisons flow into wetlands, streams and groundwater where they damage and destroy local ecology.

Fertile soil is not an inert medium but a mixture of water, air, minerals and organic matter. In most soils, minerals represent around 45% of the total volume, water and air about 25% each, and organic matter 2-5%.[114] The mineral portion consists of three distinct particle sizes classified as sand, silt or clay.

Soil health depends on the organic component that houses many living creatures along with dead material in various stages of decomposition. An acre of living soil may contain 900 pounds of earthworms, 2400 pounds of fungi, 1500 pounds of bacteria, 133 pounds of protozoa, 890 pounds of arthropods and algae, and possibly some small mammals.[115] An acre of soil may contain over 10,000 species of microorganisms, which contributes significantly to the biodiversity in living soil.[116] Unfortunately, industrial agriculture acts to kill the microorganisms with cultivation, soil compaction, chemical fertilizer, and agricultural poisons.

Soil organic matter is the smallest but most critical soil component for crops. Soil organic matter interacts to influence soil biological, chemical and physical properties and consists of raw plant residues and microorganisms, (1-10%); active organic traction, (10-40%); and resistant or stable organic matter, (40-60%) called humus.[117] Modern farmers replace the macrofertilizer, but not the humus removed by each crop.

Freedom foods are antithetical to agribusiness firms because foods grown with "natural processes" are not patentable. Similarly, bioavailable algae fertilizers and other plant inputs are natural and are not patentable. Nature was engineering marvelous products that provided for plant needs eons before Monsanto entered the business. Algae and the symbiotic microbes they attract create provide the compounds that enable plants to naturally synthesize many of the advanced compounds they need to grow and to fight disease and pest vectors. Unfortunately, U.S. government farm policy chose to support R&D on industrial food production rather than natural processes.

Farm policy

Farm policies, government sponsored research to Land Grant Universities, extension service agents, subsidies, and food support for the hungry, dictate food production. The same large agribusinesses that have addicted farmers to their branded synthetic chemistry that drives farm policy. Wealthy farmers and agribusiness like ADM, Monsanto and Cargill make enormous political donations to both parties in order to shape policy in commercial agriculture to benefit their interests. Consequently, over 99% of federal grants and R&D go to industrial agriculture. Unsurprisingly, most extension agents who are in place to help farmers and gardeners receive training in industrial production. Less than 1% of U.S. federal funding goes to organic production, R&D or training. Government funding for natural growing methods such as abundance and freedom foods rounds to zero.

India and China support natural processes R&D in food production because their leaders realize that fossil resources are finite, increasing in price, and will eventually run out. Both countries have terminated their biofuels programs with food crops for the obvious reason that food-based biofuels drive up the cost of food and the inputs to produce food.[118] China recently put a 135% tariff on their phosphorus fertilizer to insure sufficient supplies for domestic farmers. India's scientists have performed some excellent R&D with natural biofertilizers, especially focused on cyanobacteria that fix nitrogen and reduce the need for nitrogen fertilizer.

As modern farm policies evolved from 1960 to 1990, food supply and sustainability issues were not well articulated. Consumers and political leaders preferred celebrating their brilliance in designing the Green Revolution and the cheap food it provided. Leaders and policy makers ignored critical issues with GE crops, such as the need for additional cultivation, two to three times more irrigation, triple the need for fertilizer and 20 times the need for agricultural chemicals.

Few people were aware of nutrition and health issues, food security, fossil resource depletion or global warming before the 1980s. The winds of political rhetoric drowned out the few voices that challenged the fossil foods path such as Prince Charles, Vandana Shiva, Michael Pollan, Miguel Altieri, Alice Waters, and Robert Henrikson.

Today, only one third of Americans believe the scientific consensus that human actions cause global climate change. However, neither politicians nor consumers can deny that humans have caused severe fossil resource depletion and environmental pollutions with our cheap fossil food policies.

Cheap food

Freedom Foods and abundance makes little sense when the cost consumers pay for industrial foods appear to be so cheap.

Appearances can be deceiving. Consumers pay low prices for industrial foods due to government subsidies and weak math.

The U.S. government lavishly subsidizes industrial farming, big agribusinesses, big oil, water management and the fossil resources on which food production depends. For example, many farmers pay less than 2% of the true cost of irrigation water – which promotes waste and pollution. Subsidies reduce the real food cost by nearly a third, Figure 6.1. These subsidies are financed with our children's money, in government bonds held by countries like Saudi Arabia and China.

Figure 6.1 Real Cost of Food

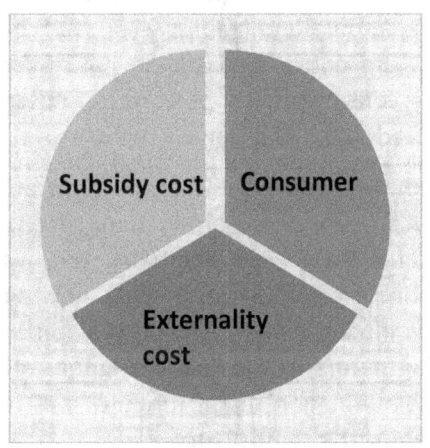

American corn subsidies decimated Haitian farmers because they could not grow food as cheap U.S. food dumped on the country as "food aid." The U.S. corn subsidies also have displaced over a 1.5 million poor Mexican farmers. Farmers were forced to leave their land because they could not compete with subsidized U.S. corn. Many of these farmers added their feet to the flow of illegal immigrants to the U.S. from Mexico. Canada, Mexico and other countries have outstanding lawsuits against U.S. subsidies with the World Trade Organization because these subsidies substantially depress the real price of food grains.

A group of more than 400 agricultural experts, known as the International Assessment of Agricultural Knowledge, Science and

Technology for Development concluded through its global and regional studies report that governments and industries need to discontinue environmentally damaging farming methods. At their 2008 meeting in Johannesburg South Africa, the group recommended "ending subsidies that encourage unsustainable practices." Political leaders in the U.S. should listen to world opinion because U.S. subsidies amplify resource consumption and pollution. Subsidies today will destroy our ability to grow our own food in the near future.

Another third of the true food cost comes from externalities such as resource depletion, environmental degradation and human health impacts, for which the food supply system fails to account. Environmental degradation alone creates about $45 billion a year in damage. No metrics are currently available for resource depletion. Neither farmers nor consumers pay a nickel for these costs. These hidden costs are shifted to our children.

A full lifecycle accounting would show fossil foods are substantially more expensive than freedom food production. Life cycle accounting includes the economic impact of degrading air, water and soil, destroying our fisheries, creating dead zones as well as cost to human and animal quality of life and health. The current generation benefits from over consuming natural resources. We ignore resource loss by failing to account for depletion in the price of our food. The next generation will not enjoy the same luxury.

Prince Charles in his Future of Food speech at Georgetown University pointed out the "curiously perverse" economic incentive system (subsidies) that too frequently directs food production. He addressed the true cost of food effectively:

> Nobody wants food prices to go up, but if it is the case that the present low price of intensively produced food in developed countries is actually an illusion, only made possible by transferring the cost of cleaning up pollution or dealing with human health problems onto other agencies, then could correcting these anomalies result in a more beneficial arena where nobody is actually worse off in net terms? It would simply be a more honest form of accounting that may make it

more desirable for producers to operate more sustainably, particularly if subsidies were redirected to benefit sustainable systems of production.[119]

Prince Charles recommends "accounting for sustainability," which represents the true cost of food production; financial costs and the costs to natural capital – the earth's resources.

When our children discover industrial agriculture lacks the natural resources to produce food, they will ask the government for increased subsidies. Unfortunately, the government will be out of funds. What country would be willing to make loans that add to the immense U.S. debt? The U.S. is already a debtor nation; we just fail to act as one.

When our children discover their fields worn out, fresh water is unavailable, fuel costs are out of reach, fertilizer mines are exhausted and agricultural chemicals have ruined their waterways – will they agree that our fossil foods were cheap?

Biofuels

In the 1990s, the Clinton administration made a critical political mistake and stopped R&D on algae for food or biofuels. Those funds were shifted to corn ethanol for biofuel. The decision by the EPA to fund a corn ethanol industry may have been the most costly decision in American history because it accelerates natural resource depletion. When the U.S. runs out of resources to produce food, who will sell us food? Where will the government find the money to buy food for hungry Americans?

The farm lobby remains so strong that corn ethanol subsidies continue at around $20 billion a year, even though ethanol consumes more fossil energy than it returns. Huge subsidies flow primarily to large agribusiness and landowners, not to family farms. Subsidies continue in spite of clear scientific proof that corn ethanol is an expensive, wasteful proposition that not only massively depletes our natural resources but creates billions of dollars in degraded and damaged ecosystems.

The 44 million acres of corn grown for ethanol in 2010 could and should be replaced by less than 2 million acres of algae biofuel production, while improving air and water quality.

Biowar I: Why Battles over Food and Fuel Lead to World Hunger (Edwards, 2007) traces the money path, primarily to one company, ADM, that initiated the biofuel industry with millions in political donations to both parties. Today ADM receives billions each year in biofuel subsidies. A biowar occurs when a country burns food, typically as a horrific act of war on another country.

In Biowar I, the U.S. became the first country to burn its own food. Biowar I ignited when the Bush Administration announced the Energy Policy Initiative in 2005, which increased biofuel subsidies and mandates, Figure 6.2. The unintended consequence of producing large amounts of corn ethanol on U.S. and world food markets was predictably higher food prices. In the eyes of the UN, World Bank and most foreign countries, the U.S. ethanol policy contributed substantially to the terrible 2008 food riots in 40 countries.

Figure 6.2 Corn Burned for Ethanol and Food Stamps

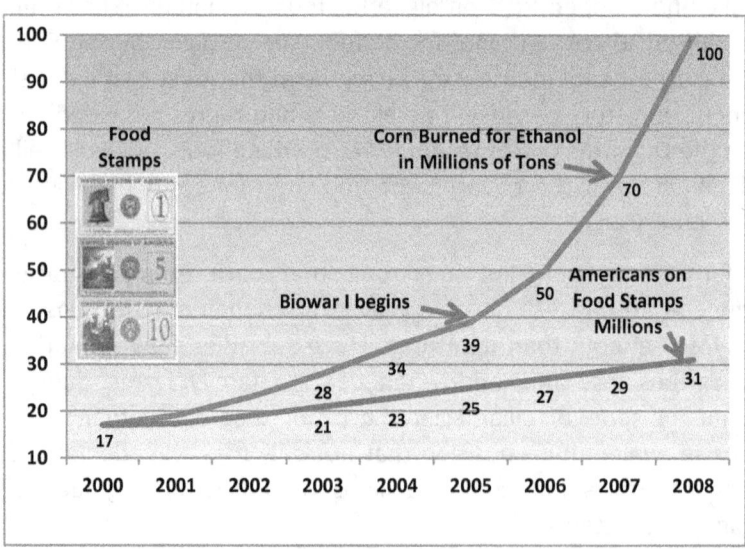

How could people in hungry countries not blame the U.S. for food shortages and price increases when prior to the ethanol program, America provided half the world's food grains and 70% of the world's corn imports? How does a country with over 60 million people receiving food support because they are hungry justify a policy of burning its citizen's food for a weak fuel additive? Over 30 million Americans are on food stamps and must abhor the concept of burning food because they know their $1 a meal buys less food. Biowar I can end by a simple government policy that states: "Subsidies should flow to sustainable, non-pollutive food and energy production systems."

In 2009, the U.S. became a net importer of food. A college sophomore could make the case that the U.S. biofuel policy is wasteful and foolish. We are burning our children's natural resources.

We may have developed a less expensive, food production system had people not had distaste to one word – algae.

Consumer behavior and algae

Why do most people have an immediate aversion to algae? The answer is false attribution. When asked to describe algae, people's top of mind typically elicits several words with strong negative connotations: "slimy, smelly, scummy and yucky." If putrefied raw meat were presented as steak, people would naturally dislike steak.

People falsely attribute the smell in ponds to algae because it certainly looks like algae. Actually, the odor comes not from algae but from the bacteria that attack and eat the algae. The bacteria consume all the oxygen algae added to the water, causing entrophication. When the aquatic organisms are deprived of oxygen, they die and begin rotting, which adds smell and slime to ponds. Healthy algae give off lots of oxygen and smell similar to walking through a redwood forest – without the redwood trees, of course.

Weed algae in ponds grow in diverse communities of many microorganisms and are different from the algae we cultivate in abundance microfarms. Gardeners know they must remove the weeds from the garden or the weeds will take over the garden and consume all the nutrients. Abundance growers control weed algae in order to enable healthy production of the target species.

The yucky factor seldom occurs with edible algae. At a recent Rotary meeting at a Kobe Steakhouse, the Rotarians were served three forms of algae at the luncheon: algae soup broth for taste and thickening, sea vegetable in the soup for texture, color and visual appeal, and separately a cold seaweed salad for color, taste and texture. The Rotarians were asked after lunch if anyone liked algae to eat, which elicited the yuck factor, as all were extremely negative. They reversed judgment when they understood sea vegetables and algae salad provided the colors, taste and texture. The Kobe manager brought out a large variety plate of sushi made with algae. The Rotarians consumed the sushi plate in one pass.

Nori represents only one of more than 1,000 sea vegetable and has a world market value of over $3 billion a year. Nori serves as a luxury food. Cooks wrap Nori around a rice ball with a slice of raw fish on the top. Toasting or baking brings out Nori's rich flavor and flakes complement rice or noodles. Epicurean cooks make a Nori soy sauce reduction for meat or seafood. Nori adds a spicy taste to jam and wine. Chinese cooks use Nori in soups and for seasoning fried foods. Many other sea and freshwater algae foods await commercialization.

Many forms of algae do not taste good by themselves. Some algae may need processing with other ingredients, similar to soybeans, to create palate-pleasing foods. Algae-based foods will become more attractive with informative food labels.

Food labels

Current food labels display only a few bits of information about calories and fat. We propose a more comprehensive food label that represents the nutritional value, sustainability, social value and ecological costs of food. Since the label is too long for many food products, we are working on a smart phone app that will allow consumers to access this label information for any food product.

We plan to expand the concept of taste tests to demonstrate the benefits of freedom foods. These blind taste tests will ask consumers first for their preferences based on standard consumer behavior protocols that include taste, aroma, texture, color, sweet, sour, salty, bitter and umami (hardy or savory). The secondary test will ask for consumer preferences for food A or B using a food label similar to Figure 6.3.

Table 6.3 Expanded Food Label
Example: One cup of Corn and one cup of Spirulina

	Industrial food	Freedom food
Nutrients		
Calories	132	
Calories from fat	16	
Total fat	1.8 g, 3%	
Saturated fat	0.3g, 1%	
Polyunsaturated fat	0.9g, 3%	
Monounsaturated fat	0.5g, 2%	
Cholesterol	0	
Sodium	23mg, 1%	
Dietary fiber	29.3g, 17%	
Sugars	5g, 1%	
Protein	5g, 1%	

Health and nutrition		
Genetically engineered, GE	90% of U.S. food grains	No
Health risk from GE	unknown	No
Nutrient dilution / empty calories	Yes	No
High in vitamins	No	Yes
High in minerals	No	Yes
High in antioxidants	No	Yes
High trace elements	No	Yes
Social justice		
Affordable to all	No	Yes
Provides food security	No	Yes

Expanded food labels will provide consumers with additional information that is currently unavailable. Some consumers will ignore the expanded labels, just as they ignore the current brief labels. The purpose of the expanded label strategy is to support the dialogue on sustainable and affordable food and energy (SAFE) production.

Freedom foods are not a panacea and offer solutions to only some food issues. A set of needed technology breakthroughs, defined in the DOE National Algal Biofuels Technology Roadmap, will be required for the optimum use of algae biomass for commercial production.[120]

Recent advances in microscopes, biotechnology, nanoscience, biophysics and bioengineering enable scientists to understand how microorganisms grow and develop. These recent breakthroughs, translated into working demonstration models, will enable diffusion of abundance. We encourage critiques, additions or insights to expanded food labels, any of the consumer behavior challenges or food policy issues at the open source web site www.AlgaeCompetition.com. The next section explores frequently asked questions.

Chapter 7. Freedom Foods FAQ

Freedom foods will have some unintended consequences.
We must actively monitor progress and insure transparency.

If growing freedom foods with abundance methods were easy, they would have been developed and diffused decades ago. We are just beginning the development of the practice, abundant agriculture 1.0. Initial abundance questions include the following.

1. **Farmers will never give up dirt farming and adopt abundance to grow freedom foods.**

Farmers do not have to abandon their fields or their favorite crops. Farmers can regenerate their fields and improve crop yields. Growers want to leave their fields better than they found them.

2. **We will not run out of fossil resources.**

Really? The question is not whether we will run out, only when. In many developing countries, fossil resources, especially productive seeds, water and fertilizer, have already become too expensive.

3. How will we pay for transforming our food system?

The freedom foods industry can begin as a supplemental food supply that does not compete with industrial foods. Policy leaders may shift subsidies from fossil foods to ecologically positive foods. Local policy leaders may begin taxing or penalizing polluters who degrade our ecosystems for this and the next generation. Leaders may also decide to shift part of the $17 billion a year for border security and $18 billion a year that goes into NASA space exploration to sustainable, healthy food. The health of our children and food security should be our national priority.

4. Algae production is still 10 years away.

No, firms have been producing algae profitably for 30 years. We conducted the only algae industry surveys to date, (for the Algal Biomass Organization and for Algae World in Asia). Those results show that over half of producers believe successful algae production is operational now or will be within three years. Over half of respondents believe that the U.S. can replace ethanol with algae in under 10 years. Over 150 companies are producing algae successfully today. Once CAPS are available for gardeners, freedom foods growth will become exponential. The Freedom Foods Revolution plans to inspire 16,000 Green Masterminds (algae producers) by 2016 who will train and 10 million by 2025.

5. Algae production is too hard.

Yes, abundance has been too difficult for nonscientists, but that is changing. I have failed three times over the last 20 years in constructing a backyard CAPS. All we need is one success – and a person or team willing to share the techniques as open source. We are engaging Green Masterminds, engineers, architects, and scientists at www.AlgaeCompetition.com to share their innovative designs and production breakthroughs. Please join the fun and become a full-fledged Green Mastermind.

We have successfully prototyped smartcultures production near fields. Crop yields and cost reductions exceeded our expectations. However, we need to make the CAPS less expensive, more portable and easier to operate and maintain. These upgrades will happen soon. Diffusion of innovation before CAPS operate reliably and efficiently would create the same false promises that doomed many prior global food initiatives.

6. **Algae production is too expensive.**

No, fossil agriculture is too expensive because consumers pay only one-third the cost of food. One-third is masked by subsidies and another third, ecological cost, is unaccounted for and ignored. A lifecycle analysis of fossil foods that accounts for all costs including extraction, depletion, waste, erosion, ecological pollution and all the impacts to human and animal health, will show that industrial agriculture is more expensive than freedom foods. The key political question will be how long people are willing to ignore the substantial subsidy and ecological costs imposed by industrial agricultural production.

7. **If freedom foods were this good, they would already available.**

No. Prior to recent biotechnology breakthroughs, freedom foods were impractical. Recent innovations in biophysics, biochemistry, bioengineering and a host of other disciplines have converged to make abundance possible now. Of course, we will have to apply new technologies to make production easy and reliable.

8. **Isn't freedom foods simply organic food production?**

Freedom foods growers use organic methods to cultivate foods low on the food chain, not traditional crops. Organic growers of land-based crops avoid, to the degree possible, fossil inputs such as GE seeds, chemical fertilizers, and agricultural poisons. However, organic production typically requires more fertile soil, freshwater and fossil fuel than industrial agriculture. Both organic and industrial growers can adopt abundance methods to improve yields, reduce costs, and operate more sustainability.

9. **There is no such thing as a free lunch.**

This is my favorite critique because it represents the food chain with the little fish eaten by successively larger fish. Freedom food production is not free in an economic sense because all inputs have some cost in terms of transportation, labor or capital. Recovering, recycling and reusing farm waste stream nutrients closes the nutrient cycle and frees farmers from continual extraction from their fields and the need to continually purchase more fossil fertilizers and chemicals. While abundance may not be free, the practice offers a very cheap lunch compared with the escalating costs of modern fossil agriculture.

10. **People will refuse to eat algae.**

True, some people will continue to eat meat and potatoes, assuming they can still afford the price of meat. Some may have cultural values that forbid ingesting microorganisms – other than the algae, fungi, bacteria, molds, mildews, yeasts and viruses that ride naturally on everything we eat. These organisms also reside in our gut, where they work symbiotically to provide bioavailable nutrients from the food we ingest.

People who want to abstain from eating algae directly may use abundance methods to grow algae for aquaculture, hydroponics or aquaponics. Others may grow algae to feed their birds, dairy or meat animals. Farmers may grow algae in the smartcultures model to improve yields and quality of their field crops.

New food processing technologies will transform empty calorie snack foods into tasty, low-fat, high nutrient health foods. A food renaissance will transform convenience foods into foods that build strong bodies and minds. Algae flour and oils will enable people to have their chocolate cake with ice cream and whipped cream – and eat it too – without fat and calorie guilt. Superb algae-based gourmet foods will become the rage in upscale restaurants. Many algae foods already grace fine restaurant menus.

11. **What if algae run amok, e.g. the movie *Solyent Green*?**

Algae are already plentiful in the natural environment. Should some algae get loose; the culture will grow to the limits of the in situ nutrients and then stop. Of course, growers can also kill algae with chlorine, as we do in pools and aquariums. A Google.com search on algae provides 10:1 information on how to kill algae rather than how to cultivate algae.

12. **What if algae carry a disease or create toxins?**

There is a legitimate concern that algae, just like any other food, may carry salmonella, e-coli or another pathogen. Continual monitoring, possibly remotely, can identify and stop cultures that may be problematic.

The research on algae toxins identifies the species and conditions necessary to produce toxins. Growers avoid both the species and conditions. Intense R&D continues on algae toxins because toxins have stopped tumor cell reproduction for 30 types of cancer.

13. **Aren't waste streams dirty?**

Yes and the pathogens can be killed using natural tools such as solar heaters. Food processors have used the technologies successfully for 50 years throughout the food supply chain.

14. **Aweehhh!**

Algae are just a bunch of cells. Yes, just like a carrot but with 10 times more beta-carotene per ounce. Carrot cells differentiate themselves into root, stems and leaves, while alga cells do not. Macroalgae and sea vegetables are made of many independent undifferentiated cells but organized as pseudo-stems and leaves, similar to land plants.

15. **Freedom foods does not address population management.**

True, and without some form of population management, many are doomed to poverty and hunger. Freedom foods addresses large families indirectly because families in developing countries often have extra children in order to assure labor for the farm. Abundance removes the need for extra children as a form of supplying farm labor because abundance enables food production with relatively light work performed by almost anyone.

16. **Will freedom foods take jobs away from farmers?**

No, abundance methods engage farmers in a new set of actions that makes farming more profitable. Many aspiring farmers can practice farming because freedom foods require modest space. Freedom food will create a new industry with millions of green jobs.

17. **How can we get started now?**

Use social media to explore your ideas. Create teams that design environmental landscapes, microfarms, algae foods and menus. Post your ideas at www.AlgaeCompetition.com. Use the site to extend and refine your concepts and find other like-minded social entrepreneurs.

18. **The farm lobby is too strong with its huge political lobbies and guaranteed subsidies.**

The farm lobby shames the U.S. the Congress and all Americans. Subsidies pay enormous amounts of public money, 85% of which goes to large agribusinesses and a few wealthy landowners – not hard working family farmers. The farm lobby may embrace freedom foods in order to answer to Congress for the health of our children, the overconsumption and loss of our precious natural resources, and the pollution of our ecosystems.

19. **No highly productive, low-cost, easy maintenance CAPS are on the market today.**

True, which means we must design, develop, demonstrate and diffuse CAPS that anyone can use.

20. **There are no businesses to support freedom food production.**

True today, but consider the many phenomenal entrepreneurial opportunities in this new industry.

21. **How could we subsidize farmers to help transition to abundance?**

America needs is an agriculture or homeland security secretary who has courage to make a smart policy statement:

> *Food and energy subsidies should benefit the production of sustainable and ecologically positive farming practices.*

Congress could then shift the $20 billion a year in subsidies and costs currently wasted on corn ethanol to multiple forms of truly sustainable energy.

22. **You do not address vegetables versus meat.**

True, we elected to leave those arguments for others. Others have written terrific books and articles on the value proposition for eating lower on the fossil food chain with vegetables. Notable works include the *Moosewood Restaurant Cooks at Home: Fast and Easy Recipes for Any Day* by the Moosewood Collective, *Quick-Fix Vegetarian: Healthy Home-Cooked Meals in 30 Minutes* by Robin Robertson, and *How to Cook Everything Vegetarian: Simple Meatless Recipes* by Mark Bittman and Alan Witschonke.

23. **You will never be able to train people to grow algae.**

We do not intend to train just anybody. We plan to train Green Masterminds, who share freedom food passion and are willing to conduct peer training with others. We need first to train farmers and master and hobby gardeners with deep knowledge about growing plants. We will need freedom foods demonstration sites in every community, so there will be many grower opportunities.

We also need to engage scientists, technologists and students with biotechnology knowledge. We need green masterminds with a passion for improving food quality and security for their family and community. We need educators who codify the training materials and convey abundance methods to people without reading or language skills. We need social, political, business and religious leaders who develop vision, values and action plans to move abundance forward. We need a green environmental groups and NGOs to promote the abundance environmental and health benefits.

We need media people who clearly articulate both the value proposition and the urgency for action. We need University professors and scientists to conduct the research that critically examines each of the value propositions for abundance. Most of all, we need our children to engage in this new food production model because their survival may depend on abundance methods and freedom foods. Critiques, additions or insights are encouraged at the open source web site www.AlgaeCompetition.com.

Books on SAFE production, including *Green Algae Strategy, Crash! Green Solar Gardens, Abundance, and Abundant Agriculture,* are available free for download in color PDF at the above site for students, faculty and food and energy policy leaders. Schools, colleges and universities globally use books in the *Green Algae Strategy* series for courses in a wide variety of disciplines, especially for our sustainable future. The *Green Algae Strategy* series is also available on Amazon.com and other retailers.

Chapter 8. What is the Food e-Footprint?

How light is a butterfly's footprint?

The four primary metrics for sustainable food and biofuels are life cycle analysis, ecological footprint, freshwater footprint, and carbon footprint. LCA, also known as ecobalance, is a technique to assess environmental impacts associated with all the stages of a product's life from-cradle-to-grave. For food and biofuels, LCA examines environmental impacts from raw material extraction through cultivation, crop inputs, harvest, food transportation, refining, processing, supply chain, and disposal or recycling. Lifecycle analysis does not measure the threat of resource extinction and does not factor in financial, physical or weather risks.

An ecological footprint creates a metric for human demand on the Earth's ecosystems. The footprint compares human demand with our planet's ecological capacity to regenerate. For 2006, humanity's total ecological footprint was estimated at 1.4 planet Earths. This means societies use ecological resources 1.4 times as fast as Earth can renew them.[121] Scientists recalculate the metric annually, with a three-year lag due to the time necessary for the UN to collect and publish the underlying statistics. The full e-footprint calculation considers housing, transportation, recreation and food consumption.

A freshwater footprint measures the consumptive use of freshwater. Water that is recycled such as household use, (except yard and garden) that is reclaimed and reused is called non-consumptive water. Freshwater used for agricultural purposes and food refining are consumptive because the water is not available for reuse.

Water Footprint

A carbon footprint measures air pollution, as a derivative of the ecological footprint. A carbon footprint creates a metric for the total set of greenhouse gas emissions caused by any human activity. For foods, the carbon footprint typically expresses the amount of carbon dioxide, or its equivalent of other GHGs, emitted from production, transportation and consumption.[122]

The food e-footprint covers a wide variety of ecological factors but does not consider the possibility of resource extinction. Food production represents a special case were each of 24 fossil resources must be available to crops precisely on time or the crop fails.[123] When farmers find one fossil resource unavailable or unaffordable, they may lose their entire crop. Unfortunately, many of the vital fossil resources face extinction, especially in specific food growing regions.

The e-footprint for food consumption provides a broader metric than air pollution but narrower than the total ecological footprint. The e-footprint for food creates a metric that reflects the natural resources required to provide a consumer with food. The food e-footprint considers production, waste, risk, transportation, and pollution. The creation of a food e-footprint provides a means to measure one's impact on the planet. Most people are not aware of their food footprint. Self-awareness provides the first and necessary step for behavior change. Footprint calculations offer policy makers a set of standards that help in formulating sustainable food policy. The food e-footprint may help political, medical and business leaders make recommendations to improve our food supply.

Food chain

People that eat high on the food chain consume large amounts of dairy and meat products and leave a large e-footprint. Vegetarians diminish their footprint far lower than meat consumers do, but still substantially higher than consumers that eat freedom foods.

Vegetarians leave a modest ecological footprint, not by choice but due to the way producers grow industrial foods. Each ton of grain consumes about 1000 tons of freshwater, as well as considerable cropland, fuels, fertilizers and chemicals. Often, 60% of food is lost in the field or the food supply chain.[124] Fossil food production creates significant pollution, carries substantial physical and economic risk to farmers, continually extracts soil nutrients, erode soil and pollute ecosystems. Industrial crops are weather intolerant, elevating the risk of crop failure.

Freedom foods avoid most of the resource consumption, pollution, risk, and weather problems that plague fossil foods. Growing freedom foods can repair air and water pollution and regenerate soils. Therefore, freedom foods consumers leave a tiny ecological footprint.

The peer-reviewed ecological footprint for food consumption remains to be constructed. The following draft model, Table 8.1, uses a hector of corn and algae for comparison. Further refinement for the model may quantify the footprint per pound of protein. The food ecological footprint provides a relative rather than absolute metric for different consumers. This descriptive model serves as an educational tool because it captures the relevant categories that make up the footprint. Another model might use the food required to supply 2,200 calories per day.

Table 8.1. Ecological Footprint for Food Consumption

Fossil inputs	Hector of corn	Hector of algae
Fertile soil – cropland	10,000 m^2 107,637 ft^2 2.47 acres	0
Fresh water	9.35 M liters 2.47 M gal	0
Fossil fuels	7.6 liters 2 gal	0.4 liters 0.1 gal
Fossil fertilizer	91 kg 200 lbs	0
Fossil chemicals	4.5 kg 10 lbs	0
Transportation	5,000 miles	5 miles
Shelf life	Days	Months
Loss to pests	15%	5%
Spoilage loss	60%	5%
Pollution		
Air – dust, CO_2, NO_x	Yes, tons	No
Poisons soil	Yes, massive	No
Poisons water	Yes, massive	No

Risk		
Crop failure	High	Low
30% yield loss	High	Low
Hard physical labor	High	Low
Fatigue	High	Low
Heavy mechanical equipment	High	Low
Risk of physical injury, disability or death	High	Low
Community health risk from pollution	High	Low
Soil		
Extracts nutrients	Yes	No
Extracts organics	Yes	No
Kills microorganisms	Yes	No
Breaks nutrient cycle	Yes	No
Degrades soil	Yes	No
Compacts soils	Yes	No
Erodes soil	Yes	No

Weather		
Yield loss; heat spike	Yes	No
Yield loss; cold spike	Yes	No
Vulnerable to drought	Yes	No
Vulnerable to wind	Yes	No
Vulnerable to rain	Yes	No
Vulnerable to storms	Yes	Moderate

The USDA reports that in 2007, U.S. cattlemen used two billion bushels (112 billion pounds) of corn to produce 22.16 billion pounds of finished grain-fed beef. Farmers used 13.3 million acres to produce the feed grains, since corn production averages about 150 bushels per acre. Each pound of beef releases about 22 pounds of CO_2-equivalent greenhouse gasses.[125] Consequently, a single year of beef production releases roughly 2.5 trillion pounds of CO_2-equivalent greenhouse gasses. Cars add about 2.7 trillion pounds of new carbon the atmosphere each year.[126]

David Pimentel calculates a steer consumes about 100 pounds of grain per pound of edible beef produced.[127] Using the basic rule that it takes about 264 gallons of freshwater to produce one pound of hay and grain, about 26,400 gallons of freshwater are required to produce a pound of beef, Figure 8.1.

Biofuels also create a substantial environmental footprint. Biofuels compete with food, accelerate fossil resource depletion, and amplify soil erosion and pollution. Each gallon of ethanol consumes 3,000 gallons of water to produce the corn feedstock.[128] When drawn from fossil aquifers that do not replenish with annual rains, the water is not available for future generations – for food or biofuels. Much of the cropland west of the Mississippi River draws irrigation water from fossil aquifers and several are predicted to run dry in a generation.

Figure 8.1 Freshwater cost of Food Grains and Beef

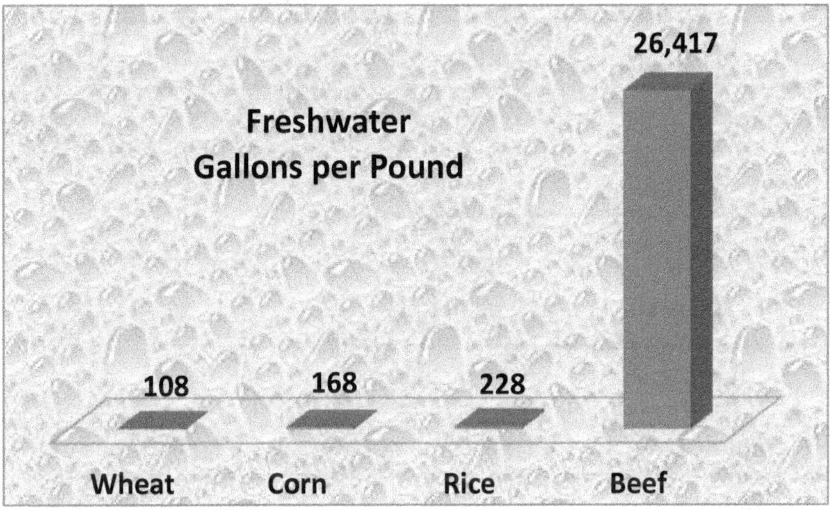

Freshwater Gallons per Pound

| Wheat | Corn | Rice | Beef |
| 108 | 168 | 228 | 26,417 |

The Renewable Fuels Association reported that in 2010, producers converted 260 billion pounds of corn into 13 billion gallons of ethanol. Ethanol has 64% of the energy of gasoline so that is 8.3 equivalent gallons of gasoline. In 2010, U.S. farmers harvested nearly 400 million tons of grain, of which 126 million tons, primarily of corn, went to ethanol fuel distilleries. Each acre of corn production releases about 4500 pounds of CO_2. Therefore, ethanol feedstock adds another 700 billion pounds of CO_2 to the atmosphere. Of course, refining the corn feedstock to ethanol and burning the ethanol in vehicles adds additional CO_2. Ethanol production is not cleantech.

Another way to create the food ecological footprint uses a rating scale for each category. A peer-reviewed ecological footprint for food and biofuel production remains to be constructed. A recent *Applied Energy* article lays the groundwork for an e-footprint calculation.[129] The draft proposal here collapses several ecological categories to create a 100 point metric. Beef resides at the top of the food chain, because beef requires a high multiple of the resources required by grains and other foods low on the food chain, Table 8.2.

Table 8.2. Food or Biofuel Ecological Footprint

Based on a 10-point scale where 10 is high consumption.	Beef Kilogram	Algae Kilogram
Fossil resource consumption		
1. Fertile soil – cropland	10	0
2. Fresh water	10	1
3. Fossil fuels	10	1
4. Fertilizers, chemicals and poisons	10	0
5. Transportation distance / spoilage	10	0
Pollution		
6. Air	10	1
7. Water	10	1
8. Soil	10	0
Risk		
9. Producers and community	10	1
10. Climate and weather tolerance	10	1
Total	**100**	**6**

People that eat high on the food chain consume large amounts of dairy and meat products and leave a large e-footprint. Vegetarians diminish their footprint far lower than meat consumers do, but still substantially higher than consumers that eat freedom foods.

Vegetarians leave a modest ecological footprint, not by choice but due to the way producers grow industrial foods. Each ton of grain consumes about 1000 tons of freshwater, as well as considerable cropland, fuels, fertilizers and chemicals. Often, 60% of food is lost in the field or the food supply chain. Industrial food production creates significant pollution, carries substantial physical and economic risk to farmers. Industrial foods continually extract soil nutrients, erode soil and pollute ecosystems. Industrial crops are weather intolerant, elevating the risk of crop failure. Freedom foods moderate the risk, and weather problems that plague fossil foods. Growing freedom foods can repair air and water pollution and regenerate soils. Therefore, freedom foods consumers leave a tiny ecological footprint.

The e-footprint for organic produce is probably around 50/100. Organic foods consume more ecological resources than industrial foods, except for fertilizers, chemicals and poisons. Some organic producers farm local to consumers and many enrich local communities.

The e-footprint for corn biofuels approaches 70/100. Corn requires intense application of natural resources. Corn production is extremely pollutive as each acre of corn releases 2.25 tons of CO_2 to the atmosphere plus nitric oxide that has 296 times the warming capacity of CO_2. A single acre of corn releases about 4,000 gallons of water each day from evapotranspiration. Corn requires ten times more nitrogen fertilizer than other food grains, which is not only consumptive but also extremely pollutive.

Fertilized soils release more than two billion tons of greenhouse gases every year, especially CO_2, methane and nitric oxide. Each cropland acre loses about 54 pounds of nitrogen, 13 pounds of phosphorus, 264 pounds of potassium and 132 pounds of calcium annually, which farmers typically replace with mined chemicals.[130]

Corn grows in rows, which amplifies soil erosion. Each acre of corn erodes over 6 tons of topsoil, in normal years. In 2011, the extreme rains and floods created a high multiple of 6 tons per acre.

Algae-based biofuels could yield an e-footprint of 20 or lower if producers use waste or brine water for nutrients and waste CO_2 for their carbon source. Growers could also avoid most fossil fuels by using renewable energy from wind, waves, geothermal or solar energy.

The food ecological footprint does not account for the substantial food health issues or the use of GE crops. Michael Pollan, *The Omnivore's Dilemma*, Jeffrey Smith, *Seeds of Deception*, and Marie-Monique Robin, *The World According to Monsanto*, cover the health and GE crop issues effectively.

Chapter 9.
The Freedom Foods Revolution

Freedom food consumers benefit from superior nutrition and taste, yet leave the ecological footprint of a butterfly.

The Freedom Foods Revolution acts to transform our food supply to fossil-free, healthy, clean, and regenerative. We collaborate to break the GE monoculture monarchy and regain our liberty to choose foods that support the Foods Bill of Rights.

Freedom foods exist today only on the drawing board. No one has cultivated an ounce of protein independent of fossil resources, which makes freedom foods and exciting challenge. Various growers have found solutions for each fossil input, but no one has put together a fossil-free food production system – yet.

The first freedom foods producers will probably be algae companies that cultivate algae using pristine brine water. They will not have to add the step of cleaning waste or ocean water. Their challenge will be to source cheap CO_2 and renewable energy, such as wind or solar to mix the culture and harvest. These early producers will probably not be able to locate near consumers.

Distributed small and medium size production offer the highest potential for local jobs and food democracy. Large production facilities concentrate wealth in the hands of a few. Large production sites also limit innovation to the relatively few people engaged in the firm. Large firms create value for their stakeholders by hoarding their intellectual property, which serves the company but not society.

In order to build a freedom foods industry, we need to design affordable growing systems usable by nearly anyone, nearly anywhere. Producers have used cultivated algae production systems, CAPS for decades, but not very efficiently. Most systems are too expensive, too large and require considerable expertise to operate. Current CAPS are not adaptable to distributed local production.

We know how to build CAPS but we need to bring the cost down by a factor of ten and make operations easier, faster and better. Please engage in our fascinating collaboratory, www.AlgaeCompetition.com with your ideas for algae landscape designs, growing systems and algae menus and foods.

Growers will need remote monitoring systems. These systems will enable growers to send questions and pictures of their culture to experts, Green Masterminds, who can help them with their needs. Monitoring systems can also build grower intelligence and a knowledge base that benefits all future growers. No practical knowledge base for non-scientists exists today for algae cultivation.

Growers will also need handbooks, online tutorials and videos to get started. These Green Mastermind training materials need to be clear and understandable for people who cannot read. Many people with the highest need for good food lack education. Simple training materials combined with remote monitoring systems will enable food production by the poor in slums or rural areas, homeless, ethnic minorities, prisoners, elderly, and other disadvantaged groups.

Objectives

Freedom foods objectives include production support and education.

Production support for microfarmers:
- Build abundance production skills for farmers, gardeners, the disabled, elderly, students, curious big kids, and prisoners.
- Share and advise on microfarm construction and operation.
- Discover and sustain a variety of species in a species bank and make them available to growers.
- Monitor cultures remotely to support growers and assure quality.
- Celebrate and promote traditional and novel freedom foods products, along with their lore and preparation.
- Construct and advise on cooperative processing facilities.
- Organize celebrations of local and regional producers and their cuisine.
- Encourage efforts to preserve family farms where microfarms may be co-located.

Educate:
- Health issues such as micronutrients, vitamins and minerals.
- Therapeutic values such as moderating obesity, diabetes and other diseases.
- Sensory issues such as taste, texture, aroma, and color.
- The significance of a positive carbon and ecological footprint.
- Benefits of distributed, scalable microfarm production locally.
- Risks of monoculture and reliance on a genomes or varieties.
- Risks of industrial agriculture, big agribusiness and factory farms.
- Differences between fossil foods and freedom foods.
- Risks and resource consumption from genetic engineering.
- Risks associated with dependence on genetic monocultures.
- Risks of residues from pesticides, herbicides, fungicides and other agricultural chemicals.
- Encourage ethical buying in local marketplaces.

We want to accomplish these objectives with social networks and open source collaboration.

Sustainability goals

We hope to achieve our freedom foods sustainability goals by 2025 that focus on enhancements in health, social equity, environment and economics. These goals are achievable by creating the capacity for distributed freedom foods production globally so growers can grow foods and coproducts for their family and community needs locally.

We plan to accomplish these goals by training a foundation cadre of 16,000 Green Masterminds (freedom foods growers) by 2016 that will help train 30 million Green Masterminds by 2025.

Goals such as #1, reduce obesity and diabetes by 50%, are not practical by 2025. Our plan is to make a substantial reduction of specific diseases in target regions. Once we demonstrate the efficacy of the freedom foods approach, those efforts can migrate to larger regions. No one knows today whether we can eliminate these diseases with freedom foods in combination with other therapeutics, or cut their occurrence by 50%. We chose the stretch goal to demonstrate what we believe is possible, with focused effort.

The Freedom Foods Revolution offers many significant outcomes. Possibly the most significant are health, new jobs and transportation fuel savings. Radically reducing obesity and diabetes may be a top priority for the many countries afflicted. Resolving the disease will save tens of billions in health care and educational costs and restore positive lifestyle for millions of families.

The Freedom Foods Revolution may not resolve some of the target diseases. We know that current cognitive and medical approaches have not slowed the increase in diabetes or several other Western diseases. A systemic new healthy foods model offers a novel solution that warrants medical, scientific and public debate.

Table 9.1 Freedom Foods Sustainability Goals by 2025

Area	Freedom Foods Sustainability Goals
	With affordable and accessible freedom foods.
Health	1. Reduce obesity and diabetes 50%.
	2. Reduce ALS, Lou Gehrig's disease.
	3. Eliminate blindness from vitamin A deficiency.
	4. Eliminate goiter and stunting from lack of iodine.
	5. Control 30 expressions of cancer.
	6. Reduce digestion problems and diseases 50%.
	7. Reduce air pollution diseases 50%.
	8. Reduce Western diseases 25%.
	9. Reduce water pollution diseases 50%.
	10. Reduce food insecurity for U.S. children 75%.
	11. Resolve food deserts by 80%.
	12. Add fresh produce to food banks 50%.
	13. Reduce global nutrient deficiencies 50%.
	14. Reduce U.S. nutrient deficiencies 90%.
	15. Improve nutrient density per bite 100%.
	16. Expand food diversity 1000%.
	17. Enhance sensory appeal 33%.
	18. Reduce farmer poison exposure 33%.
	19. Reduce rural community exposure 66%.
	20. Reduce poison residue on produce 100%.
	21. Reduce GE crops 80%.
	Assumes 50% of food grown locally.
Social – local, freedom	22. Enable distributed local producers 1000%. (Net: 10 times more good local jobs.)
	23. Improve local growers' incomes 100%.

food production	24. Reduce food transportation costs 50%.
	25. Save 9 billion gallons of food transportation fuel.
	26. Diminish GHG in cities 50%.
	27. Decrease black soot particulate pollution 66%.
	28. Cut food distance 66%.
	29. Reduce water treatment cost 25%.
	30. Reduce smoke stack pollution 66%.
	31. Cut food waste and spoilage 50%.
	32. Enable food production by the elderly, the disabled, homeless, prisoners, schools, colleges, other institutions and special needs farms.

Annual savings.

Environment – resource preservation and pollution avoidance	33. Save 40 million acres of prime cropland.
	34. Save 24 million tons of soil erosion.
	35. Save 3 trillion gallons of freshwater.
	36. Save 5 billion gallons of diesel fuel on farms.
	37. Save 3 million tons of chemical fertilizers.
	38. Save 2 million tons of agricultural poisons.
	39. Save billions of birds, bees, bats, farm animals and aquatic creatures.
	40. Reduce dead zones by 66%.
	41. Improve waters fit for human recreation by 50%.
	42. Improve fishable waters by 66%.

	Annual savings.
Economics	43. Save $20 billion in diabetes health costs.
	44. Save $20 billion in special education costs.
	45. Save $40 billion in Western diseases costs.
	46. Save 50% of $43 billion from soil erosion.
	47. Save 50% of $125 billion from air pollution.
	48. Save 50% of $300 billion from water pollution.
	49. Save $36 billion in food transportation.
	50. Save $20 billion food loss and spoilage.

Transforming food production to a distributed model with many producers close to consumers will accomplish most the social equity, environment and economics goals. Local production will create tens of thousands of new green jobs.

Placing microfarms in prisons can take a huge cost, waste management, and turn it into a profit center, while providing green training and jobs for inmates. America is the only country in the world with over twice as many prisoners as farmers. Prisons offer a great opportunity to educate green values while producing food. Special need farms have proven that food cultivation offers significant therapeutic value for a wide range of mental disorders commonly found among prisoners. The therapeutic value of freedom foods in prisons may exceed the economic value.

Food transportation cost reduction creates a valuable benefit set. The local freedom foods model with 50% of the food grown within 50 miles of consumers would save five times more liquid transportation fuel than the entire ethanol program. It would also save the 46 million acres of prime cropland planted in corn feedstock. Removing 50% of food trucks from our highways will substantially reduce pollution, road maintenance, and trucks that crush cars and people. A 50% reduction in black soot particulates in cities will reduce congestion, smog, and respiratory diseases significantly.

Local production reduces the substantial costs associated with packaging, preservatives, storage, transport, shrinkage and spoilage. Consumers will benefit from fresh local foods. Local foods reduce truck traffic in cities, creating cleaner city environments. Unlike the medical solutions that require further R&D, the local freedom foods local production model is not theoretical but practical. Not only do local foods save trillions of gallons of fuel, it preserves natural resources for our children. The local freedom foods model offers novel solutions that deserve medical, scientific and public debate.

The freedom foods sustainability goals will need public policy support in order to change from the tyranny of the present. Vested interests, big agribusiness, will take substantial losses in a new food production model. However, only a few people monopolize farm policy today while many people are victims of their reach for profits.

Please engage in the Freedom Foods Revolution. Help us regain our liberty to choose good, healthy food while saving our environment.

Policy leaders

We need to ask food policy leaders who are appointed to assure our security and rights to explain how their policies provide value. Suggested questions include the following.

1. How your policies support the food bill of rights?
2. How long will each of the critical inputs to fossil foods be available and affordable to farmers?
3. How can the biofuel policy focused on corn ethanol continue when practically all the principle goals have proven to be false?
4. What are the human, animal and ecosystem health and sustainability effects from GE crops?
5. Why do crop subsidies continue despite World Court lawsuits by our neighbors and global allies?
6. How much do crop subsidies accelerate fossil resource extinction, U.S. debt, and environmental pollution?
7. How can one department, agriculture, equitably administer both crop subsidies and food support?

8. What is the freshwater cost of every gallon of ethanol west of the 100th parallel? How many ethanol acres use irrigated cropland? The current U.S.GS number used for policy decision is 10%. This number is wrong by at least 100%, unless the ethanol plant maps are mistaken.[131]

9. What will you tell our children about:
   - The rising cost of food?
   - The unfortunate health effects from GE monocultures?
   - Food security, when 1/5 Americans are already on food support?
   - Food costs for fossil foods built on an eroding foundation of non-renewable resources?
   - Industrial foods and their health?
   - GE foods and their health?
   - GE foods and risk from a single pest vector?
   - The extinction of their fossil resources required for industrial food production?
   - Their severely degraded and polluted ecosystems?

10. Does our current industrial agricultural policy help or hurt our health, human services, and educational systems?

Policy leaders should consider appointing a food czar, independent of the USDA, to lead a scientific, consumer and farmer committee that lays out a sustainable food, health and human services policy.

Our path forward

Society needs to eject from the death spiral imposed by fossil foods: continually rising costs, water scarcity, soil degradation, severe waste and pollution – followed eventually by crop failure. Society will benefit from replacing GE foods with naturally biodiverse foods that are clean, nutrient dense and healthy. We also need a food supply that produces independent of climate chaos.

Freedom foods produced with abundance methods offer the first new agricultural technology in 60 years. Scientists and food companies have known about the productivity and nutritional benefits of cultivating microorganisms for over 100 years. However, technology barriers and subsidized industrial foods blocked this new industry.

Breakthroughs in biotechnology, biophysics and bioengineering make freedom foods practical today. Over 150 companies are growing algae and other microorganisms today but primarily for feed, fertilizer and biofuels. These producers follow the industrial model, leveraging fossil resources to maximize production. Next generation producers will grow freedom foods that liberate growers and consumers from their dependence on fossil resources.

What will our children think of our cheap foods policy when they have no resources to produce food and ecosystems too degraded to support production? The resource extraction and ecological costs will be disastrous. We need a new food system for our children.

Subsidies disguise the true cost of food and should be re-evaluated. Subsidies that accelerate extraction, waste and ecological degradation should be eliminated. Those subsidies may shift to production systems that repair and enhance ecosystems.

Policy leaders should begin accounting and possibly taxing resource extraction and pollution. Until consumers understand the consequences of artificially cheap food, they will continue to make fossil choices – especially when they have no alternative.

Some unanticipated issues are certain to arise in abundance production. Therefore, we need monitoring systems that are transparent and accessible. Mobile phones with cameras make monitoring systems possible and practical.

The design of affordable and easily operational CAPS remains our primary challenge. Fortunately, collaborative social networks like www.AlgaeCompetition.com provide a practical solution. We can pool our knowledge and share it globally. Our collective actions will enable people to grow food and other forms of energy for their family and community locally.

Acknowledgements

Thanks first to my best friend and life partner, Ann Ewen, who made this project possible. Ann supported both the "ah's" and the "aha!'s"

This product would not have been possible without the extraordinary research of David and Marcia Pimentel, Lester Brown, President of the Earth Policy Institute and Jeffery Sachs of the Earth Institute at Columbia University. Professors Qiang Hu, Milton Sommerfeld and Bruce Rittman from Arizona State University supported questions on molecular biology and algal production. Environmental scientists Al Darzins, Eric Jarvis and Mike Siebert at NREL were very helpful with renewable energy sources. Thanks also to great advisors who elevated *Abundance* from a solo to an orchestra. Adriana Delgado and Chia-wen Tsai assisted with proofing. Any remaining typos are mine.

Science	Business – Econ.	Agribusiness
• Ben Cloud	• Robert Henrikson	• Jon Ewen
• Dan Childers	• Mark Allen	• Gary Wood
• James Elser	• Alan Resnik	• Doug Young
• Carol Johnston	• Gary Dyer	• Jim Robertson
• Greg O'Reilly	• Herb Roskind	• Tracy Penwell
• Andy Ayers	• William Cockayne	• Barry Spiker

Also helpful were the published works of Paul Ehrlich, Sandra Postel, Nobel Laureate Al Gore, Harvey Blatt, Fred Pearce, Michael Pollen, Brian Halweil, Clay Jason and Linda Graham. High-content websites were a great support such as Algaebase, U.N., W.H.O., the National Resources Defense Council, Sierra Club, Green Peace, Audubon Society, Union of Concerned Scientists, Center for Energy and Climate Solutions, Clean Water Network and Public Citizen. Also useful were government sources including DOE, EPA, U.S. DA, NOAA and NREL.

Mark Edwards

Mark designs nutritious, sustainable and affordable food and energy (SAFE) production systems available to all people on Earth. Mark pursues abundance; to create food security for all and to help growers leave every field better than they found it.

Mark graduated from the U.S. Naval Academy in engineering, oceanography and meteorology, where Jacques Cousteau motivated and mentored his interest in the oceans and global stewardship. He holds an MBA and PhD in marketing and consumer behavior and has taught agribusiness, food marketing, leadership, sustainability and entrepreneurship at Arizona State University for over 30 years.

Mark served as marketing director for the Longevity Research Institute directed by Nathan Pritikin. The LRI focused on actions designed to improve the diet and exercise behaviors for people with health needs. The work led to the Pritikin diet and to Pritikin health foods. Mark also served as a director for a Fortune 50 foods company and has done extensive R&D on new foods, sources and consumer behavior. He has consulted for Monsanto, DuPont, Nabisco, Quaker Oats, General Mills, Borden, Coca-Cola, Frito-Lay, Disney, GE, Intel, J&J, Merck, GM, Bank of America, and many other companies.

Mark served as CEO of TEAMS Intl. for 24 years, the software and assessment firm he founded based on his research on advanced assessment technologies, talent and leadership assessment and new product development. He served as lead consultant for more than 400 firms globally. He has worked with senior executives at 15 large U.S. oil and gas firms as well as British Petroleum and Saudi Aramco. Many U.S. departments and the military retained him, including DOE, DOD, Special Forces and the U.S. National Labs.

The Green Algae Strategy Series

Mark R. Edwards

The Green Algae Strategy Series focuses on creating Sustainable and Affordable Food and Energy – "SAFE" production. **The Green Algae Strategy Series** are available for free downloading in color PDF for students, teachers and food and energy policy leaders at www.AlgaeCompetition.com. They are also available on Amazon.com and other retailers. Teachers, professors and policy leaders use these SAFE production books in schools and colleges globally for courses in sustainability, engineering, business, politics, social entrepreneurship, food, water, energy, ecology, environment and world future.

BioWar I: Why Battles Over Food and Fuel Lead to World Hunger, 2007. BioWar I, where food is burned for fuel, must be ended by withdrawal – not of soldiers, but of damaging agricultural subsidies.

Green Algae Strategy: Engineer Sustainable Food and Fuel. 2008. Algae offer solutions for sustainable and affordable food and energy because algae are the most productive biomass source on Earth. *Best Science Book* **– 2009, Independent Publisher Awards.**

Green Solar Gardens: Algae's Promise to End Hunger, 2009. Algaculture in small but beautiful solar gardens and algae microfarms distributed globally will enable SAFE production locally.

Crash: The Demise of Fossil Foods and the Rise of Abundance. 2010. Traditional fossil based agriculture sits precariously on a foundation of unsustainable fossil resources that will become unaffordable and then will run out. Abundant agriculture is sustainable because it uses plentiful inputs that are cheap and will not run out.

Smartcultures: Nature's tiny Genius – Algae – Reverses Pollution and Regenerates Degraded Ecosystems, 2011. Farmers may recycle farm wastes to their fields using abundance microfarms. Smartcultures give 20 – 30% higher yields by providing bioavailable nutrients at just the right time. Farmers save 30 – 40% by reducing input costs and reduce ecological pollution by 90%.

Abundance: Sustainable Fossil-free Foods with superior Nutrition and Taste; less Pollution and Waste, 2010. Abundance presents the value proposition for algae food, feeds and other forms of energy using plentiful resources that will not run out. Abundance growers can clean the air and water while they grow foods with superior nutrient density and better sensory values, including color, texture and taste.

The tiny Plant that saved our Planet. The incredible true story of Tiny, Mighty Al. Tiny Mighty Al saved our planet by eating the bad carbon genie, which enabled the earth to cool and gave us oxygen. Al saved us again by becoming the bottom of the food chain and providing all living creatures with nutritious food. If we educate our children, maybe they will prompt us to take action — now. **Winner of the 2011 Nautilus Silver Medal for the Best Children's book.**

Abundant Agriculture: Smartcultures enable superior Nutrition and Yields from Regenerated Fields, 2010. Abundant agriculture represents the first new form of agriculture in 60 years. Abundant agriculture produces sustainable food, feed, fiber and other coproducts using primarily non-fossil resources that are plentiful, affordable and often surplus. Abundant agriculture growers use abundance methods to produce healthy, nutritional foods.

Freedom Foods: Superior Nutrition and Taste from low on the Food Chain for People, Producers and Our Planet, 2011. Freedom foods liberate consumers to make healthier food choices. Freedom foods are sustainable and grow free of fossil resources, GMO material, and other agricultural chemicals and pesticides. Freedom foods growers use abundance methods that do not compete with industrial agricultural production, yet produce foods superior in nutralence and health value. Freedom foods enable consumers and growers to eat healthy, eat hearty, and leave the ecological footprint of a butterfly.

Index

References

[1] Edwards, Mark. *Abundance: Sustainable Fossil-free Foods with superior Nutrition and Taste; less Pollution and Waste*, Tempe: CreateSpace, 2010.

[2] Edwards, Mark R. Green Algae Strategy: Engineer Sustainable Food and Fuel, CreateSpace, 2008, 44.

[3] Warner, Jennifer. CDC: Kids Lack Access to Healthy Food Choices, WebMD Health News , April 26, 2011.

[4] Center for Food Safety, http://www.centerforfoodsafety.org /2011/03/18/

[5] http://www.cdc.gov/diabetes/pubs/pdf/ndfs_2011.pdf

[6] Goodland, Robert and Jeff Anhang. Livestock and Climate Change, *World Watch*, November/December, 2009.

[7] Pimentel, David and Marsha Pimentel. *Food, energy and society*, 3rd edition, New York: CRC Press, 2008, 27.

[8] http://www.ers.usda.gov/briefing/organic/Farmsector.htm

[9] Thelen, Kurt D. What is the Direct Carbon Footprint of Biofuel Relative to Gasoline? Michigan State University, Crop & Soil Sciences, https://www.msu.edu/~thelenk3/Acrobat/C%20footprint.pdf

[10] Marder, Jenny. Farm Runoff in Mississippi River Floodwater Fuels Dead Zone in Gulf, NPR, *Science*, May 18, 2011.

[11] Edwards, *Abundance*, 12.

[12] Edwards, Mark R. *Abundant Agriculture*, Tempe: CreateSpace 2010.

[13] Cui-Hua Qi, Min Chen, Jie, Soong, Bao-Shan Wang. Increase in aquaporin activity is involved in leaf succulence of the euhalophyte suaeda salsa, under salinity. *Plant Science*; 176;2, pp. 200-205, Feb 2009.

---

[14] Nutralence is a new word that refers to a plant that naturally stores nutrients. High nutralence produce is nutrient dense – the opposite of empty calories.

[15] Jeff Norrie, Seaweed Research. *American Fruit Grower*, 128;3, pp. 48-50, Mar 2008.

[16] Ibid.

[17] Luescher-mattli, M. (2003), Current Medical Chemistry-Anti-Inflammatory Agents., 2, 219-225.

[18] MacArtain P, Gill CIR, Brooks M, Campbell R, Rowland IR. Nutritional value of edible seaweeds. *Nutrition Review*, 2007, 65:535-543.

[19] Garcia-Casal MN, Pereira AC, Leets I, Ramirez J, Quiroga MF. High iron content and bioavailability in humans from four species of marine algae. *Journal of Nutrition*, 2007, 137:2691-2695.

[20] Kavaler, Lucy. *Green Magic: Algae Rediscovered*. Thomas Crowell, NY, 1983, 99-101.

[21] Spolaore P, Joannis-Cassan C, Duran E, Isambert A. Commercial applications of microalgae. 2006 *J Biosci Bioeng* 101:87-96.

[22] Garcia-Casal, et. al. 2007.

[23] Dhargalkar, V. K. and X. N. Verlecar (2009). "Southern Ocean seaweeds: A resource for exploration in food and drugs." Aquaculture 287(3/4): 229-242.

[24] Lisheng, L. et. al. Inhibitive effect and mechanism of polysaccharide of spirulina on transplanted tumor cells in mice. Marine Sciences, Qindao, China. N.5, 1991, p.33-38.

[25] Luescher-mattli, M. (2003), Current Medical Chemistry-Anti-Inflammatory Agents., 2, 219-225.

[26] Palmquist RE. 2008. Apparent response to homotoxicology, salmon oil and blue-green algae in a single geriatric canine case of episodic mentation changes. JAHVMA. April-June 27 (1): 10-15.

[27] McCarty MF, Barroso-Aranda J, Contreras F. NADPH oxidase mediates glucolipotoxicity-induced beta cell dysfunction--clinical implications. Med Hypotheses. 2010 Mar;74(3):596-600.

[28] Lee EH, Park JE, Choi YJ, Huh KB, Kim WY. A randomized study to establish the effects of spirulina in type 2 diabetes mellitus patients. Nutr Res Pract. 2008 Winter;2(4):295-300.

[29] Muthuraman P, Senthilkumar R, Srikumar K. Alterations in beta-islets of Langerhans in alloxan-induced diabetic rats by marine Spirulina platensis. J Enzyme Inhib Med Chem. 2009 Dec;24(6):1253-6.

[30] Garbuzova-Davis, Svitlana. *The Open Tissue Engineering and Regenerative Medicine Journal*, 2011, (3:36-41).

[31] Gupta S, Hrishikeshvan HJ, Sehajpal PK. Spirulina protects against rosiglitazone induced osteoporosis in insulin resistance rats. Diabetes Res Clin Pract. 2010 Jan;87(1):38-43.

[32] Shytle DR, Tan J, Ehrhart J, Smith AJ, Sanberg CD, Sanberg PR, Anderson J, Bickford PC. Effects of blue-green algae extracts on the proliferation of human adult stem cells in vitro: a preliminary study. Med Sci Monit. 2010 Jan;16(1):BR1-5.

[33] Milton K. 2003. The critical role played by animal source foods in human (*Homo*) evolution. *J. Nutr.* 133:3886–92S.

[34] Edwards, Mark R. Algae 101 Part 26: Did Algae's Great Taste Make us do it? Algae Industry Magazine, May, 2011.

[35] Kulshreshtha, Archana et. al. Spiralina in healthcare management, current pharmaceutical biotechnology, 2008, nine, 400 – 405.

[36] MacArtain P, Gill CIR, Brooks M, Campbell R, Rowland IR. 2007 Nutritional value of edible seaweeds. Nutr Rev 65:535-543.

[37] Yamada Y, Miyoshi T, Tanada S, Imaki M.(1991) Digestibility and energy availability of Wakame (Undaria pinnatifida) seaweed in Japanese. Jap J Hygiene 46;788-793.

[38] Edwards, Mark R. *Abundance*, 88.

[39] Yamamoto S, Tomoe M, Toyama K, Kawai M, Uneyama H. 2009 Can dietary supplementation of monosodium glutamate improve the heatlh of the elderly? Am J Clin Nutr. 90:844S-849S.

[40] Edwards, Mark R. *Abundance*, 91.

[41] Yangthong M, Hutakilok-Towatana N, Phromkunthong W. 2009 Antioxidant activities of four edible seaweeds from the southern coast of Thailand. Plant Foods Hum Nutr 64:218-223.

[42] Abad MJ, Bedoya LM, Bermejo P. 2008 Natural marine anti-inflammatory products. Mini Rev Med Chem 8:740-754.

[43] Edwards, Mark R. Smartcultures: Nature's tiny Genius – Algae – Reverses Environmental Pollution and Regenerates Degraded Ecosystems, CreateSpace, 2011, 45.

[44] Prince Charles, Future of Food Conference, Georgetown University May 4, 2011. http://washingtonpostlive.com/conferences/food

[45] Ibid.

[46] Edwards, Mark R. *Crash! The Demise of fossil Foods and the Rise of Abundance*, Tempe: CreateSpace, 2009, 54.

[47] Edwards, Mark R. *BioWar I: Why Battles over Food and Fuel Lead to World Hunger*, Tempe: LuLu Press, 2007, 185.

[48] Shiva, Vandana. The Suicide economy of corporate globalization, April 5, 2004, Znet. http://www.countercurrents.org/glo-shiva050404.htm

[49] http://www.agcensus.usda.gov/Publications/1997/Agricultural_ Economics_and_Land_Ownership/indexintro.asp

[50] Zwerdling, Daniel. India's Farming 'Revolution' Heading For Collapse, NPR, April 13, 2009. http://www.npr.org/templates/story/story.php?storyId=102893816

[51] "OECD Highlights Chinese Pollution." Financial Times, 17 July 2007. http://www.ft.com/cms/s/932c36ca-348c-11dc-8c78-0000779fd2ac.html.

[52] Brennan, Morgan. America's Most Polluted Cities, Forbes, 04.28.11.

[53] Lawrence, Felicity . Not on the Label Penguin. 2004, 213

[54] Hidden Hunger, http://www.micronutrient.org/english/View.asp?x=573

[55] Engel SM, Wetmur J, Chen J, Zhu C, Barr DB, Canfield RL, et al. 2011. Prenatal Exposure to Organophosphates, Paraoxonase 1, and Cognitive Development in Childhood. Environmental Health Perspectives: doi:10.1289/ehp.1003183.

[56] Ibid.

[57] Trasande, Leonardo and Yinghua Liu. Reducing the Staggering Costs Of Environmental Disease In Children, Estimated At $76.6 Billion In 2008, *Health Affairs*, April 2011.

[58] GE crops and pesticide use, http://www.ucsusa.org/food_and_agriculture/science_and_impacts/impacts_genetic_engineering/genetically-engineered-crops.html

[59] http://news.uns.purdue.edu/html4ever/0012.Huber.deficiency.html

[60] Pimentel, David. Food, energy and society, 75.

[61] Young, A. Agroforestry for soil conservation. Wallingford, UK: CAB, 1989.

[62] Wilkinson, Bruce H. and Brandon J. McElroy, The impact of humans on continental erosion and sedimentation, Geological Society of America Bulletin, January 2007; v. 119; no. 1-2; p. 140-156; DOI: 10.1130/B25899.1

[63] Natural Resources Conservation Service, Soil Erosion, 2006.

[64] EPA, Assessed Waters, in 2000 National Water Quality Inventory, EPA-841-R-02-001, EPA, Office of Water, Washington, D.C., August 2002.

[65] Worldwatch Report #179: *Mitigating Climate Change through Food and Land Use*, June, 2009.

[66] USDA. 2007 Natural Resources Inventory, http://www.nrcs.usda.gov/technical/NRI/2007/nri07erosion.html

[67] Diaz, Robert J. and Rutger Rosenberg. Spreading Dead Zones and Consequences for Marine Ecosystems, Science 15 August 2008: 926-929.

[68] World Hunger Statistics 2010, http://www.worldhunger.org

[69] http://www.bread.org/hunger/global/

[70] USDA, Food stamps make America stronger, 2010,
http://www.fns.usda.gov/

[71] Morgan Stanley, Little Minds need big meals, *Wall Street Journal*, Dec 15,
20011, A1.

[72] Feed America, http://feedingamerica.org/SiteFiles/child-economy-
study.pdf

[73] CDC data and statistics on obesity and diabetes in America.
http://www.cdc.gov/features/dsObesityDiabetes/

[74] http://www.cdc.gov/diabetes/pubs/pdf/ndfs_2011.pdf

[75] Conkin, Paul K. A Revolution down on the farm: The transformation of
American Agriculture since 1929, U. of Kentucky Press, 2008, 164.

[76] Environmental Working Group, Crop Subsidies Data Base,
http://farm.ewg.org/

[77] Cook, Ken. Government's continued bailout of corporate agriculture,
http://farm.ewg.org/summary.php

[78] Food Planet, Food summit blames trade barriers, biofuels, June 4, 2008,
www.planetark.com/dailynewsstory.cfm/newsid/48626/story.htm

[79] Edwards, Mark R. *BioWar I*, 187.

[80] Joffe-Walt, Chana. Why U.S. subsidies Brazil's cotton crop, Nov 9, 2010.
http://www.npr.org/blogs/money/2011/01/26/131192182/cotton

[81] Brown, Lester. World Grain Harvest Falling Short by 54 M Tons: Water
Shortages Contributing. Earth Policy Institute, 23 Nov. 2003, 3.

[82] Klein, Gary. The Water-Energy-Greenhouse Gas Connection, California
Energy Commission
www.bayareavision.org/initiatives/PDFs/GreeningInfill_120607_Klein.pdf

[83] Environmental Working Group, 2009. http://farm.ewg.org/

[84] Heller, Martin C., and Gregory A. Keoleian. *Life Cycle-Based Sustainability
Indicators for Assessment of the U.S. Food System*. Ann Arbor, MI: Center
for Sustainable Systems, University of Michigan, 2000: 40.

[85] Pimentel, D., Hepperly, P., Hanson, J., Douds, D., and R. Seidel. 2005. Environmental, energetic, and economic comparisons of Organic and Conventional farming systems. Bioscience 55(7):573-582.

[86] Resources for the Future, What Do the Damages Caused by U.S. Air Pollution Cost? December 17, 2007, http://www.rff.org

[87] Neuman, William. High Prices Sow Seeds of Erosion, *New York Times*, April 12, 2011.

[88] Iowa Daily Erosion Project, 2011, http://wepp.mesonet.agron.iastate.edu/index.phtml

[89] Edwards, Mark R. *Abundance*, 183.

[90] Edwards, Mark R. *Smartcultures: Nature's tiny Genius – Algae – Reverses Environmental Pollution*, Tempe: CreateSpace, 73.

[91] http://www.parkseed.com/gardening/PD/9261/

[92] www.cooksgarden.com

[93] Edwards, Mark R. *Abundance*, 192.

[94] Perry, Ann. Algae: A Mean, Green Cleaning Machine, *Agricultural Research Magazine*, 58:5, May/June 2010.

[95] Kulshreshtha, Archana et. al. Spiralina in healthcare management, Current Pharmaceutical Biotechnology, 2008, 9, 400 – 405.

[96] MacArtain P, Gill CIR, Brooks M, Campbell R, Rowland IR. 2007 Nutritional value of edible seaweeds. *Nutrition Review*, 65:535-543.

[97] Tokusoglu, O., Unal, M.K. "Biomass Nutrient Profiles of Three Microalgae: Spirulina platensis, Chlorella vulgaris, and Isochrisis galbana." Journal of Food Science. 68, 4, 2003.

[98] Lindemann B. A taste for umami. 2000 Nature Neruosci 3:99-100.

[99] Edwards, Mark R. *Smartcultures*, 22.

[100] http://www.unmillenniumproject.org/

[101] Graham, Linda and Lee Wilcox. *Algae: Biology and Biotechnology*, New Jersey: Prentice Hall, 2000: 8.

[102] Union of Concerned Scientists, Industrial Agriculture: Policy, 2009.

[103] Cimitile, Matthew. Crops absorb livestock antibiotics, science shows. *Environmental Health News*, January 6, 2009.

[104] Edwards, Mark R. *Smartcultures*, 53.

[105] Edwards, Mark R. *Abundant Agriculture*, 22.

[106] http://www.dietresearch.com/

[107] Goyal, SK. A profile of algal biofertilizer. in *Biotechnology of Biofertilizers*, Kannaiyan, S. Ed., Delhi: Narosa Publishing House, 2002, 250 – 258.

[108] Megasun.bch.umontreal.ca/protists/gallery.html algaebase.org/links/ utex.org; ccmp.bigelow.org; http://www.ccap.ac.uk; marine.csiro.au/microalgae; wdcm.nig.ac.jp/hpcc.html).

[109] Graham, Linda and Lee Wilcox. *Algae*. New Jersey: Prentice Hall, 2000: 8.

[110] Graham, L.E. The origin of the life cycle of land plants. *American Scientist*, 1985, 73; 78 – 96.

[111] Hu, Qiang. "Environmental Effects on Cell Composition." *Handbook of Microalgal Culture: Biotechnology and Applied Phycology*. Ed. Amos Richmond. Oxford, England: Blackwell Science, Ltd., 2004: 83-94.

[112] Edwards, Mark R. *Smartcultures*, 22.

[113] Vaidyanathan, Gayathri. Genetic Engineering No Match for Evolution of Weed Resistance, *Scientific American*, April 14, 2010.

[114] Tisdale, S. L. and W. L. Nelson. *Soil Fertility and Fertilizers*. 3rd ed. New York: Macmillan, 1975.

[115] Brady, N. C. *The Nature and Properties of Soils*. New York: Macmillan Publishing Co., 1974.

[116] Gliessman, Stephen R. Agroecology: The Ecology of Sustainable Food Systems, Second Edition, CRC Press; 2 ed., 2006.

[117] Plaster, E. J. *Soil Science and Management*. 3rd ed. Albany: Delmar Publishers, 1996.

[118] Edwards, Mark R. BioWar I, 84.

[119] Prince Charles, Future of Food Conference, Georgetown University May 4, 2011. http://washingtonpostlive.com/conferences/food

[120] DOE, National Algal Technology Roadmap, 2010. http://www1.eere.energy.gov/biomass/pdfs/algal_biofuels_roadmap.pdf

[121] United Nations Environment Programme, UNEP reports. http://www.footprintstandards.org

[122] Brown, Marilyn A., Frank Southworth, and Andrea Sarzynski. Shrinking The Carbon Footprint of Metropolitan America. Brookings Institution Metropolitan Policy Program, 23 Feb. 2011.

[123] Edwards, Mark R. *Crash! The Demise of fossil Foods and the Rise of Abundance*, Tempe: CreateSpace, 2009, 26.

[124] Hawkesworth, Sophie et. al. Feeding the world healthily: the challenge of measuring the effects of agriculture on health. Phil. Trans. R. Soc. B. September 27, 2010 365:3083-3097;doi:10.1098/rstb.2010.0122

[125] Avery A, Avery D 2008. Beef Production and Greenhouse Gas Emissions. Environmental Health Perspectives 116:A374-A375. doi:10.1289/ehp.11716

[126] Thelen, Kurt D. What is the Direct Carbon Footprint of Biofuel Relative to Gasoline? Michigan State University, Crop & Soil Sciences, https://www.msu.edu/~thelenk3/Acrobat/C%20footprint.pdf

[127] Pimentel, David and Pimentel, Marcia. Food, Energy and Society, 3rd Ed., New York, CRC Press, 2007, 201.

[128] Edwards, Mark R. *Green Algae Strategy: Engineer Sustainable Food and Fuel*, CreateSpace, 2008, 144.

[129] Subhadra, Bobban G. and Mark R. Edwards, Coproduct market analysis and water footprint of simulated commercial algal biorefineries, *Applied Energy* 88 (2011) 3515–3523

[130] http://www.nrcs.usda.gov/technical/NRI/2007/nri07erosion.html

[131] Edwards, Mark R. *Crash! 84.*